Principles
of
Seed Pathology

Volume II

Authors
Vijendra K. Agarwal, Ph.D.
Associate Professor
Department of Plant Pathology
G. B. Pant University of Agriculture and Technology
Pantnagar (Nainital), India

James B. Sinclair, Ph.D.
Professor
Department of Plant Pathology
College of Agriculture
University of Illinois at Urbana-Champaign
Urbana, Illinois

CRC Press
Taylor & Francis Group
Boca Raton London New York

CRC Press is an imprint of the
Taylor & Francis Group, an **informa** business

First published 1987 by CRC Press
Taylor & Francis Group
6000 Broken Sound Parkway NW, Suite 300
Boca Raton, FL 33487-2742

Reissued 2018 by CRC Press

© 1987 by Taylor & Francis
CRC Press is an imprint of Taylor & Francis Group, an Informa business

No claim to original U.S. Government works
This book contains information obtained from authentic and highly regarded sources. Reasonable efforts have
been made to publish reliable data and information, but the author and publisher cannot assume responsibility
for the validity of all materials or the consequences of their use. The authors and publishers have attempted to
trace the copyright holders of all material reproduced in this publication and apologize to copyright holders if
permission to publish in this form has not been obtained. If any copyright material has not been acknowledged
please write and let us know so we may rectify in any future reprint.

Except as permitted under U.S. Copyright Law, no part of this book may be reprinted, reproduced, transmitted,
or utilized in any form by any electronic, mechanical, or other means, now known or hereafter invented,
including photocopying, microfilming, and recording, or in any information storage or retrieval system,
without written permission from the publishers.

For permission to photocopy or use material electronically from this work, please access www. copyright.com
(http://www.copyright.com/) or contact the Copyright Clearance Center, Inc. (CCC), 222 Rosewood Drive,
Danvers, MA 01923, 978-750-8400. CCC is a not-for-profit organiza-tion that provides licenses and
registration for a variety of users. For organizations that have been granted a photocopy license by the CCC,
a separate system of payment has been arranged.

Trademark Notice: Product or corporate names may be trademarks or registered trademarks, and are used only
for identification and explanation without intent to infringe.

A Library of Congress record exists under LC control number: 86006147

Publisher's Note The publisher has gone to great lengths to ensure the quality of this reprint but points out that
some imperfections in the original copies may be apparent.

Disclaimer The publisher has made every effort to trace copyright holders and welcomes correspondence from
those they have been unable to contact.

ISBN 13: 978-1-138-50592-6 (hbk)
ISBN 13: 978-1-138-56142-7 (pbk)
ISBN 13: 978-0-203-71081-4 (ebk)

Visit the Taylor & Francis Web site at http://www.taylorandfrancis.com and the CRC Press Web site at
http://www.crcpress.com

PREFACE

The science of seed pathology is relatively young, having its beginnings in seed health testing and control of seedborne pathogens. Since the late 1970s there has been a worldwide increase in research, outreach, and training activities relating to seed pathology. Seedborne pathogens have had special consideration in seed production areas and in plant quarantine activities. Recognition of the increased interest and importance of this branch of plant pathology was given by the creation of the Danish Government Institute of Seed Pathology for Developing Countries, Copenhagen, in 1967, the Seed Pathology Committee by the American Phytopathological Society in 1976, and the International Society of Plant Pathology in 1977.

This book was written to serve those interested in seed pathology. It is designed to serve as a textbook as well as a reference book for students, teachers, and researchers, and for seed health testing, seed production, and plant quarantine personnel. It is to be used as a guide to the literature. Much of the illustrative material has come from the authors' files used for teaching or from their own research. Teachers will want to supplement this book with examples from their own experience and research or with information and data from other seed pathology programs.

The authors hope that this book, in addition to being of value to seed and plant pathologists, will be useful to agriculturalists interested in crop production. It was written in part to stimulate research in seed pathology and its importance to the role of seedborne inoculum in the epidemiology and control of plant diseases.

The authors wish to thank G. B. Pant University of Agriculture and Technology (GBPUAT) and the University of Illinois at Urbana-Champaign (UIUC) for the use of library facilities in preparation of the manuscript. Thanks are given to M. L. Verma (GBPUAT) and Janice A. Draper (UIUC) for typing the initial and subsequent drafts, respectively, of the manuscript. The authors are indebted to Aliza Halfon-Meiri, Department of Seed Research, The Volcani Center, Bet-Dagan, Israel; and Indra K. Kunwar, Visiting Research Associate (UIUC), for helpful suggestions, advice, and proofreading of the manuscript. Drawings are by Lenore Gray.

V. K. Agarwal wishes to thank GBPUAT for permission to take up the task of preparing the material for this book and is especially grateful to Y. L. Nene, formerly from GBPUAT and now at the International Crops Research Institute for the Semi-Arid Tropics, Andhra Pradesh, India, for his guidance, inspiration, and encouragement throughout the author's career. He owes special regard and gratitude to his grandfather, Babu Ram Prasad (deceased) and to his parents for their guidance and inspiration, and to his wife, Kiran, and his daughters, Priyanka and Sheelu, for their patience and cooperation during the preparation of the manuscript.

THE AUTHORS

Vijendra K. Agarwal, Ph.D., is Associate Professor of Plant Pathology in the Department of Plant Pathology, College of Agriculture, Govind Ballabh Pant University of Agriculture and Technology (GBPUAT), Pantnagar, U.P., India.

Dr. Agarwal obtained his B.Sc.(Honours) Agriculture and Animal Husbandry in 1965, the M.Sc.(Agriculture) in Plant Pathology in 1967, and his Ph.D. in Plant Pathology in 1975 under the guidance of Dr. Y. L. Nene from GBPUAT. He joined the Department of Plant Pathology in July 1967 and served as Senior Research Assistant, Assistant Professor, and Associate Professor. He worked as Research Scholar at the Danish Government Institute of Seed Pathology for developing countries, Copenhagen, from January 1969 to February 1970.

Dr. Agarwal has taught basic courses in plant pathology and seed pathology at the graduate and postgraduate levels. He has developed research and teaching programs in seed pathology at GBPUAT. His research has concentrated on standardization of techniques for the detection, seed transmission, seed certification, and control of seedborne pathogens. He has published over 55 research papers, 2 books, and 60 other research abstracts and technical articles all in the area of seed pathology. Dr. Agarwal has attended and presented invitational lectures at several national and international conferences, seminars, workshops, and the like.

James B. Sinclair, Ph.D., is Professor of Plant Pathology in the Department of Plant Pathology, College of Agriculture, University of Illinois at Urbana-Champaign, Urbana (UIUC). Professor Sinclair received his B.Sc. degree from Lawrence University, Appleton, Wisconsin in 1951 and his Ph.D. in plant pathology from the University of Wisconsin, Madison in 1955 under J. C. Walker with whom he continued to work with a postdoctoral appointment until 1956, when he accepted a position in the Department of Plant Pathology, Louisiana State University, Baton Rouge (LSU). At LSU he served as an Assistant Professor, Associate Professor, and then Professor until 1968. Also, he was an Administrative Assistant to the Chancellor from 1966 to 1968. He joined the Department of Plant Pathology, UIUC in 1968 as a professor of international plant pathology. He was campus, then all-university coordinator for the Illinois-Tehran Research Unit, 1974—1978.

Professor Sinclair has taught five graduate courses in plant pathology; has planned, participated in and given invitational lectures at numerous national and international conferences and workshops; has worked in over 40 countries professionally; and has directed the research of 56 graduate students, of whom 15 have completed a portion of their thesis research at an overseas institution. He is a member of many national and international professional organizations. Professor Sinclair formed and was chairman of the Inter-American Seed Pathology Group and initiated the formation of seed pathology committees in the American Phytopathological Society and International Society of Plant Pathology. He served from 1979 to 1983 as the Chairman, Seed Pathology Committee, International Society of Plant Pathology.

Professor Sinclair's research has been primarily on seed- and soilborne pathogens of soybeans and other crops and their control, and on the uptake and translocation of systemic fungicides in various crop plants. He has done pioneer research in the pathology of soybean seeds.

He has published over 188 refereed research papers; 196 research abstracts; and authored, edited or co-edited 17 books and 198 other articles. The most recent recognition for his accomplishments was the presentation of the ICI/American Soybean Association Research Recognition Award in 1983 and the Paul A. Funk Award in 1984.

TABLE OF CONTENTS

Volume I

Volume II

Chapter 9

FACTORS AFFECTING SEED TRANSMISSION

Approximately 1500 seedborne microorganisms and viruses, many of which are plant pathogens, have been recorded on about 600 genera of crop plants.[1] Successful establishment of seedborne inoculum in the field is an important factor in the epidemiology of many pathogens. Organisms may be seedborne but not seed transmitted. Seed transmission often is complex, depending on several interactions. A number of factors influence successful seed transmission and establishment of infection in a subsequent crop.

I. CROP SPECIES

The establishment of seedborne infection in the field depends upon the host cultivar. Seeds of a resistant cultivar may not be infected, or even if infected, seed transmission may not occur.

The lack of seed transmission of the loose smut of wheat pathogen, *Ustilago tritici*, in certain cultivars may be due to embryo resistance, a noncompatible reaction with the pathogen resulting in seedling death, or mature plant resistance.[2,6] Embryos of the wheat cultivar Keystone are susceptible to *U. tritici* hyphae and seeds normally are infected, but mature plants are resistant.[7] Similarly, embryo infection of the barley cultivar Emir is not transmitted to the plant.[8] Resistance to *U. tritici* in wheat cultivars can be expressed when seedlings outgrow smut fungus hyphae.[5] In the barley cultivar Jet, embryos can be infected, but an incompatible reaction generally prevents development of infected shoots; however, if infection results in stunted seedlings, they may partially recover and produce healthy tillers.[9] Partial earhead infection of barley or wheat by the loose smut fungi can be due to the fungus not keeping pace with ear development.[10] In rye cultivars susceptible to *Urocystis occulta*, hyphae development is rapid and abundant, whereas in resistant cultivars, it is restricted and does not spread into meristematic regions.[11]

Resistance to transmission of *Ustilago avenae* in oats is expressed as necrosis of epidermal cell walls and restriction of hyphal growth. However, depending upon the level of resistance, the host reacts in a variety of ways. In resistant lines, hyphae may not penetrate the cell wall. In cultivars with less resistance, penetration is achieved, and in 7 days necrosis develops in the surrounding host tissues and the hyphae die; or hyphae are found after 7 days in the seedling coleoptile and mesocotyl, but degenerate in less than 21 days; or hyphae are abundant, penetrate deeper into host tissues and, after 21 days, remain as hyphae in the mesocotyl but do not invade the growing point and meristem; or in susceptible cultivars the growing point and meristem are invaded.[12]

In initial stages of infection, hyphae of *U. hordei* develop equally in barley coleoptile tissues regardless of the degree of cultivar resistance. Hyphae degeneration begins after 5 to 6 days, with the number of degenerated hyphae corresponding to the degree of resistance. Cell nuclei react to hyphal penetration by increasing in size. Vacuolization of protoplasm and lysis of the hyphal walls also are observed.[13]

The spore load of *Tilletia caries* that causes appreciable (9.5%) bunt in wheat in the susceptible cultivar Jenkins Club gives a smut-free crop in the resistant cultivar, Marquis.[14] *T. caries* penetrates both resistant and susceptible wheat seedlings, but it does not develop beyond penetration and development of hyphae within epidermal cells of resistant cultivars. In cultivars with less resistance, the fungus progresses into deeper parts of coleoptile and sheath tissues of the earliest true leaves about as readily as in a susceptible cultivar until seedlings are 9 days old, after which it is restricted. In susceptible cultivars it develops in the young leaf blades, nodes, internodes, and growing points.[15]

Pseudomonas syringae pv. *glycinea* (bacterial blight of soybean) develops larger populations on germinating soybean seeds of a susceptible cultivar than a resistant one, thus demonstrating pre-emergence host-pathogen specificity.[16] Seed transmission of a particular virus varies with the species or cultivar of a plant, as well as with the virus or virus strain. For example, seed transmission of BCMV varies among bean cultivars from 1 to 75%,[17,18] BSMV among barley cultivars from 15.6 to 64.6%,[19] alfalfa mosaic virus among alfalfa cultivars from 0.6 to 10.3%,[20] and lettuce mosaic virus among lettuce cultivars from 1 to 8%.[21] Seed transmission of alfalfa mosaic virus in alfalfa varies from 15.5 to 27.5% depending upon cultivar.[22] A 3% seed transmission of tobacco ringspot virus (strain 98) was found in lettuce seeds of cultivar Paris Island Cos, but none was found in seeds of the cultivar Imperial 615.[23] Grogan et al.[24] reported 5.1% seed transmission of squash mosaic virus through seeds of zucchini squash, whereas in seeds of Early Summer Golden and crookneck squash cultivars there was no transmission. Similarly, Kennedy and Cooper[25] reported 20.6% seed transmission of SMV in seeds of the soybean cultivar Harosoy but none in Merit. No seed transmission of peanut mottle virus occurred in four cultivars of large-seeded peanuts (Early Runner, Florigiant, Florunner, and Virginia Bunch 67) but there was transmission in four small-seeded cultivars (Argentine, Spancross, Starr, and Tifspan).[26]

The degree of seed transmission can vary among individual plants within cultivars. Seeds from individual soybean plants of the same cultivar showed a range of 0 to 35% seed transmission of SMV[27] while plants infected with tobacco ringspot virus showed 100% seed transmission.[28] Peanut seeds from 7 of 30 plants produced virus-free seeds and the other 23 produced seeds with peanut mottle virus ranging from 0.5 to 8.3%.[26]

Reasons for variation in seed transmission among cultivars may be the genetic makeup of the cultivars involved or differences in severity of strains on cultivars. The barley cultivar Hypana has a lower level of seed transmission of BSMV than the cultivar Atlas in part due to the greater genetic variation in Atlas.[29] The interaction between BSMV and various barley cultivars fall into three categories: (1) very susceptible with little or no seed produced; (2) tolerant, in which some strains of the virus survive indefinitely; and (3) resistant, in which plants either do not become infected or, if infected, seed transmission is low or absent. Thus, seed transmission depends upon the nature of the cultivar and survivability of the virus strain.[30]

II. ENVIRONMENT

Seed transmission of pathogens and their establishment and development in the host is influenced by environmental conditions, with moisture and temperature being the most important factors. These factors also affect seed germination, spore germination, the infection process, and subsequent inoculum spread.

A. Moisture

Atmosphere and soil moisture play an important role in seedling infection and subsequent disease establishment and spread. Soil moisture influences spore germination and viability as well as seed emergence. For example, the occurrence of loose smut of oats, caused by *Ustilago avenae,* is higher at 30%, lower at 60%, and lowest at 80% water-holding capacity.[31] Excessive moisture reduces oxygen and thus inhibits spore germination. Similarly, the incidence of *U. hordei* in oats at 20% water-holding capacity and 20°C can be as high as 94%, while at 60% water-holding capacity it is 48%.[32]

Similarly, seedling sorghum infection by *Sphacelotheca sorghi* is favored by low soil moisture and temperature.[33] In contrast, covered smut of barley (*U. hordei*) is favored by higher soil moisture (50% water-holding capacity) rather than low (40%).[34] Dry soil favors infection and development of *Urocystis occulta* in rye.[35] Soil moisture has a variable effect on bunt of wheat, caused by *Tilletia caries;* the percentage infection was 55, 22, and 11 at 40, 20, and 80% water-holding capacity, respectively.[36]

Low soil moisture (48.8% water-holding capacity) promotes spore germination of *T. caries* and *T. foetida* and delays development of susceptible wheat seedlings, and at higher moistures (85% moisture-holding capacity) seedling infection is least.[37] A combination of 10°C and an 11, 13, 18, and 24% soil moisture capacity is favorable for wheat bunt (*T. caries*) but infection drops off rapidly at 10% and is absent at 9%. At all moisture levels infection was greater at 10°C and intermediate at 15 and 5°C, with none developing at 25°C.[38] The incidence of pre-emergence death of wheat seedlings due to *Fusarium avenaceum, F. culmorum, F. nivale,*[39,41] or *Septoria nodorum*[42] is higher in dry than wet soils. High soil moisture inhibits seed germination and development of *Rhynchosporium secalis* hyphae thus preventing infection of barley seeds during germination.[43]

B. Temperature

Temperature affects spore germination, the infection process, and disease development. The optimum temperature for infection of wheat plants by seedborne *T. caries* and *T. foetida* ranges from 5 to 10°C.[37,38,44] In this range, germinating spores are highly infective and wheat-seed germinating is slow, thus only few plants escape infection. A range of 15 to 20°C favors wheat-seed germination more than spore germination; when spores and seeds germinate at 25°C, many plants escape infection. The escape mechanism is not known, but is attributed to rapid growth of the host. Wheat plants are susceptible to infection at 15 to 25°C, provided that the fungus spores germinate in the soil and are in an infective stage.[45] A range of 10 to 25°C and 15 to 20°C is optimum for infection of barley by *Ustilago hordei* and *U. nigra*, respectively.[46,47] Infection of wheat by *Urocystis agropyri* is optimum at 10 to 20°C and low at 5°C, with no infection taking place at 25°C regardless of soil moisture.[48] The two barley cultivars, one susceptible, Odessa, and one moderately resistant, Persicum, to the covered smut fungus (*Ustilago hordei*) show parallel reactions at the same temperature. Susceptible barley cultivars develop over 95% covered smut at 12°C soil and 16°C atmospheric temperature or 20°C soil and 24°C atmospheric temperature and at diurnal alternations of the two temperature regimes. Resistant cultivars develop no symptoms at 12°C soil and 16°C atmospheric temperature, with a low percentage of sori, primarily of inflorescens, occurring at the other temperature regimes.[46]

With loose smut of wheat (*U. tritici*), smutted earheads, earheads with reduced teliospore formation, and slender green heads with few to no teliospores are produced at 18.3, 23.8, and 29.8°C, respectively.[49] Maximum development of *U. tritici* races C_1 and C_3 occur on the wheat cultivar Kota at 23°C, whereas at 15 and 20°C the incidence of loose smut is reduced, and at 6°C the reduction is highly significant. At the lower temperatures hyphae in ears often fail to sporulate.[50]

For *Sphacelotheca sorghi* to develop in sorghum, the mean maximum soil temperature during the infection period should be at or below 24°C and above −1°C.[51,52] A maximum temperature range between 19 to 20 ± 1°C and a minimum of 8 to 10 ± 1°C followed by intermittent rains during wheat anthesis is favorable for *Neovossia indica* infection.[52] Wheat seeds infected with *N. indica* (10% in 1970—71 and 1.5% in 1971—72) failed to establish field infection due to unfavorable environmental conditions.[54]

Transmission of *Drechslera graminea* in barley seeds increases between 12 to 15°C and decreases or is prevented above 15°C in field plantings of naturally infected seeds.[55] Soybean seeds encrusted with *Peronospora manshurica* oospores (Figure 18) show systemic infection at 13°C soil temperature in 40% of the seedlings due to slow seed germination, whereas at 18°C and above, none of the seedlings show systemic infection.[56] Maize kernels infected with *D. maydis* result in 1 and 8% wilted seedlings after 3 weeks at 13 and 22°C, respectively.[57] Safflower seedlings infected by *Puccinia carthami* were 96.1, 76.2, 67.3, and 29.3% at 5, 10, 15, and 20°C, respectively.[58]

Seedling infection from seeds with *Sclerospora graminicola* hyphae occurs by planting pearlmillet seeds for 12 hr under artificial daylight at 23 to 25°C.[59] Disease development caused by *F. nivale* in wheat occurs at the highest levels at low temperatures in dry soil and least in warm dry soil.[41] The percentage of infected barley seedlings caused by *Xanthomonas campestris* pv. *translucens* var. *cerealis* is 6 at 10°C and 77 at 35°C.[60]

Seed transmission of some viruses is influenced by the environment, with temperature being a major factor. Crowley[61] reported no transmission of BCMV in seeds of bean plants grown at 17 to 19°C, whereas 16 to 25% transmission was obtained from seeds produced at 20°C. Singh et al.[62] found 3% seed transmission of BSMV in barley line C.I.5020, and none in C.I.3212, C.I.3212-I, or C.I.4219 when plants were grown at 16°C, but when plants were grown at 24°C, seed transmission in the latter three cultivars ranged from 7 to 28%. At 20.8 to 39°C under intense light, no seed transmission of BCMV occurred in urdbean, but at 20 to 30°C under diffuse light, 10% seed transmission was detected.[63] Transmission of different nepoviruses through *Stellaria media* seeds is affected differently by ambient temperature during seed production. Raspberry ringspot and tomato black ring viruses are seed transmitted at 14, 18, or 22°C; arabis mosaic virus at 14°C; and strawberry latent ringspot and tomato black-ring viruses at 22°C.[64]

C. Wind-Blown Rain

Wind-blown rain is essential for the spread of seedborne inoculum and initiation of disease such as bacterial blight of soybeans (*Pseudomonas syringae* pv. *glycinea*).[65] Secondary spread of halo blight (*P. syringae* pv. *phaseolicola*) was recorded after hail storms but not rain storms in Idaho in 1966.[66] The absence of rainfall during the growing season in California precludes spread of *P. syringae* pv. *lachrymans* in cucumber-seed fields.[67]

D. Light

Increased levels of bunt occur in the wheat cultivars Hope and Marquis with increased daylength. Hope plants developed 64.1, 17.5, and 0.8% bunt; and Marquis plants 32.7, 17.8, and 1.9%, exposed to light for 24, 10 to 11, and 8 hr daily, respectively.[68] The wheat cultivar Canus is resistant to three races of *T. foetida* and five of *T. caries* under natural-day conditions, but under continuous light, a breakdown in resistance to certain races is observed. Ulka wheat cultivar is susceptible to all but one race under both daylength conditions, but resistance to this race is reduced under long days.[69] Systemic transmission of *Peronospora parasitica* in cabbage seedlings occurs at the cotyledon stage with <16 hr/day.[70] Systemic symptoms of seedborne viruses often are masked under high light intensity.

III. INOCULUM

Seed transmission can be influenced by the amount of inoculum as well as the type, virulence, and location of inoculum in seeds.

A. Minimum Effective Level of Inoculum for Seed Transmission and Establishment in Seedlings or Plants

A spore load of 36,000 to 150,000 per seed is required to produce the maximum infection of bunt *(Tilletia caries)* in wheat.[14] The incidence of *T. caries* in a less-susceptible cultivar, Austrobankut, and a susceptible one, Stamm 101, is proportional to the spore load (3 to 3500 spores per seed).[71] The incidence of *Fusarium nivale* in wheat tends to increase with an increase in the spore load up to 50,000.[41] A heavier spore load of *F. avenaceum* is required for disease development than of *F. culmorum.*[72] Rice grains with few conidia grouped at one infection site without hyphae, few conidia scattered all over the surface of the seed coat without hyphae, or light to heavy sporulation along with light to profuse hyphal growth of *Drechslera oryzae* result in 46, 80, and 92 to 100% losses in seedlings, respectively (Figure 30).[73] *Ascochyta pinodella* and *Mycosphaerella pinodes* cause greater yield reductions than *A. pisi* because lower seed infection by *A. pinodella* and *M. pinodes* can cause greater yield loss. Higher levels of *A. pisi* in seeds is necessary to cause similar losses in yield.[74] Transmission of *Pseudomonas solanacearum* in *Capsicum* seeds occurs at an infestation level of 1000 but not 50 propagules per seed.[75]

B. Inoculum Location

Popp[76] reported that infection of wheat plumule buds by *Ustilago tritici* was correlated with adult plant infection, but Khanzada et al.[77] did not observe plumule bud infection and found that scutellar infection of wheat embryo produced infection in adult plants. *Alternaria brassicicola* is externally and internally seedborne in Brassicas, with seedling infection correlated with the latter.[78] Infection of oats by *U. avenae* occurs more frequently in plants from infected inner grains of the second flower in a spikelet than outer grains of the first flower. Spores adhering to the glume exterior are incapable of causing infection because seedlings cannot be infected from spores in such a position until hyphae have traversed the glume length, a distance of about 1 cm. Spores on the glume generally are incapable of forming sufficiently long hyphae to reach seedlings. Spores within glumes are more favorably situated to cause infection.[79]

C. Type of Inoculum

Beet seeds contaminated by *Uromyces betae* teliospores may give rise to spermogonia on hypocotyls originating from basidiospore infection; thus, teliospore-contaminated seed clusters are a potential risk for introduction of *U. betae* into new areas. However, the risk of beet rust spread by seedborne uredospores is less, because beet-seed germination requires a longer time than uredospore germination. Thus, it is possible that plants will not be available for infection when the uredospores germinate.[80] Seedlings from safflower seeds covered with uredospores of *Puccinia carthami* show no symptoms after 1 month, while those grown from seeds covered with teliospores are 90% rusted.[58]

Different strains of the same virus can differ in their degree of seed transmission. Grogan and Schnathorst[23] found that strain 98 of tobacco ringspot virus was transmitted up to 3% in lettuce seeds of Paris Island Cos whereas a calico strain was not. Transmission of bean yellow mosaic virus in cowpea varies with the strain, ranging from 0 to 55%.[81] The bean-infecting strain of SBMV is not seed transmitted in cowpea[82,83] while the strain infecting cowpea is transmitted. The serological group IA of squash mosaic virus is seed transmissible in pumpkin cantaloupe (*Cucumis melo* var. *reticulatus*), honeydew melon (*C. melo* var. *inodorus*), pumpkin (*Cucurbita pepo*), and scalloped summer squash (*Citrullus vulgaris*), but group IIA is transmitted only in pumpkin and squash seeds.[84] Similarly, a strain of squash mosaic virus, SMV-W, is seed transmitted in watermelon but not the SMV-C strain.[85] Four isolates of peanut

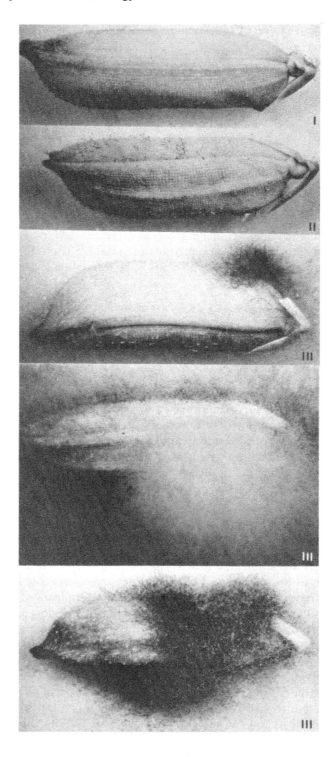

FIGURE 30.   Categories (I—III) of infection by *Drechslera oryzae* on rice (paddy) (*Oryzae sativa*) seeds incubated on blotters for 7 days at 20°C under 12/12 hr alternating cycles of near UV light and darkness. (From Aulakh, K. S., Mathur, S. B., and Neergaard, P., *Seed Sci. Technol.*, 2, 385, 1974. With permission.)

mottle virus differ in frequency of seed transmission in the peanut cultivar Starr: M_1 = 0.3%, M_2 = 0%, M_3 = 8.5%, and N = 0%. Isolate M_2 is not seed transmitted in large-seeded peanuts but is at a very low frequency (0.23%) in small-seeded peanuts.[26] The tobacco streak virus isolate A-TSV is seed transmitted in soybean, but W-TSV isolate from tobacco is not.[86] Maximum earcockle infection of wheat was obtained with two nematode galls (approximately 2×10^4 larvae per 1000 g soil) and any increase in inoculum caused a reduction in infection due to competition for food.[87]

IV. SURVIVAL OF INOCULUM

Seedborne pathogens must survive various stages of seed development and storage for successful transmission. Seed transmission of pathogen can vary from country to country and region to region depending upon weather and storage conditions. Seedborne pathogens, in general, survive longer in temperate than tropical regions. Survivability of seedborne inoculum is discussed in Chapter 7.

V. CULTURAL PRACTICES

Cultural practices apparently affect establishment of seedborne infection in the field, but available data are sparse and inadequate.

A. Soil Type
Soil type affects seed-germination rate and seedling growth. Seed transmission of loose smut in wheat and barley is higher in a clay-sand mixture than in heavier soils.[47] Symptoms of *Septoria nodorum* in wheat are more marked on seedlings grown in loam than on those in chalky soil with a high moisture content.[88] *Tilletia caries* and *T. foetida* infection in wheat in a susceptible cultivar is affected by soil type and temperature during the infection period. There is no appreciable difference in infection level in seedlings of Marquis and Thatcher cultivars in Hempstead silt loam and Mendon loam soil at 10 and 15°C, but at 5°C Marquis develop 72.5 and 30.3% bunt in Mendon loam and Hempstead silt loam soil, respectively.[89]

B. Soil Reaction
Information on the effect of soil pH on seed transmission is scanty. Neutral soils generally favor disease development. Seed transmission of *Ustilago hordei* (covered smut of oats) varied from 4 to 12%, 64 to 92%, and 8 to 18% at pH 4.6, 7.4, and 8.6, respectively.[32] Similarly, seed transmission of *T. caries* in wheat was 42% at pH 7.9 compared to 6% at pH 5.6.[44] In contrast, seed transmission of *U. hordei* in barley is favored by acid soil, with the transmission rate approximately double of that found in alkaline soils.[34] In rye, the highest level of *Urocystis occulta* was recorded when seeds were planted in soil at pH 7.36.[35] The severity of *Fusarium nivale* in wheat increased with increases in pH.[41]

C. Seeding Rate
Seeding rate can influence tiller infection. In barley, the amount of smutted tillers decreased by half to four times with increased seeding rate.[90] The density of sowing had no effect on the percentage of infected plants, except on smutted ears. Any condition that increases tillering lowers the percentage of diseased ears.[91]

D. Depth of Sowing
Deep sowing and cool temperatures, which tend to slow germination and plant growth, increase frequency of *T. caries*.[92,93] These conditions lengthen the period of

Table 11
EFFECT OF DEPTH OF PLACEMENT
OF *ANGUINA TRITICI* GALLS IN
SOIL ON THE INCIDENCE OF EAR
COCKLE OF WHEAT[87]

Depth (cm)[a]	Ear cockle infection (%)	No. of tillers/ pot	No. of galls/ pot	Grain yield/ ear (g)
2	74.4	9.3	87.6	0.4
4	35.5	8.0	48.3	0.7
6	25.3	6.7	13.3	1.0
8	16.6	4.0	2.3	1.0

[a]   Galls and seeds placed at same depth.

seedling susceptibility. Tiemann[94] found that a low rate of seed transmission of loose smut of barley was associated with a moderate depth of planting compared to either shallow or deep seeding. Pedersen[95] found that seed transmission of loose smut in barley decreased with an increase in sowing depth — 40 and 20% at 4- and 12-cm depth of sowing, respectively. The incidence of dwarf bunt in wheat was highest when seeds were planted at or near the soil surface. As the depth of seeding increased up to 10 cm, the infection percentage decreased.[96] As the depth of sowing increased from 2.5 to 15 cm, the disease index of *F. nivale* in wheat increased.[41] Deep sowing increased infection of *T. caries* in winter wheat and was correlated with delayed seedling maturation.[93] Maximum ear cockle infection in wheat occurred when nematode galls were placed with seeds at a 2-cm depth,[87] with an infection decrease if galls were placed deeper than the seeds because larvae fail to reach the seedlings (Table 11).[97]

E. Sowing Time

The emergence of loose-smutted ears in barley extends over a period of 3 to 4 weeks. Late sowing has no effect on the number of diseased ears but may increase their rate of emergence.[98] Seed transmission of BSMV is higher in spring-seeded barley than when the same cultivars are sown in the fall.[99]

F. Fertilizers

Gassner and Kirchhoff[100] reported that seed transmission of the loose-smut pathogen in barley is less when optimum fertilizer levels are used compared to low fertilizer levels. In contrast, Pedersen[95] found that the fertilizer levels had no effect on the seed transmission. Increased nitrogen fertilizer results in increased incidence of *S. nodorum* in wheat seeds,[101] but reduced *F. moniliforme* infection in maize kernels.[102] The incidence of *N. horrida* in rice is highest in fields recently dressed with nitrogen and planted with long-grain rather than short- or medium-grain cultivars.[103]

VI. SEED ABNORMALITIES

The association of various pathogens with abnormal seeds, i.e., discolored, distorted, shrivelled seeds, reduced seed size, etc. is discussed in Chapter 12. Abnormal seeds may give rise to a higher level of seed transmission compared to normal seeds. Transmission of the loose-smut pathogens of barley and wheat is higher in small compared to large seeds.[90,94,104-107] The percent transmission of loose smut of barley was 3.2, 8.7, and 13.5 in large, medium, and small seeds with 1000-grain weights of 45.1, 33.6, and 24.5 g, respectively.[108]

Large barley seeds have less of the loose-smut pathogen than small ones. The average percentage of smutted plants ranged from 8.2 for large (retained on a 2.58-mm slotted screen) to 26.7 for the small seeds (passed in a 2.38-mm screen, retained on a 1.98-mm screen). Field germination was lowest for the small seeds, with large seeds producing vigorous, tolerant seedlings of rapid growth.[104]

Loose-smut fungus infection in small seeds of dwarf wheat cultivars (Sonora-64 and Sonalika) is higher than in large seeds. A negative correlation between seed weight and loose-smut fungus infection was recorded. In a tall cultivar, Agra Local, loose-smut fungus infection in large seeds was significantly higher than in small ones. A high positive correlation was obtained between seed weight and seed infection. A lower percentage of seed transmission than seed infection may occur using the embryo-count method since smaller seeds are removed during processing.[109]

Small seeds have a higher level of seed transmission of viruses than large ones. The rate of seed transmission of peanut mottle virus in peanut is 3.7% in small seeds (<6.00 mm) compared to 0 to 0.9% in large seeds (>6.5 to 7.9 mm).[110] Seed transmission of peanut stunt virus in peanut is low enough in seeds that are large enough for planting, and seeds that transmit the virus tend to be low in germination and give rise to weak seedlings. Thus, seed transmission of the virus under field conditions is unlikely to occur.[111]

Seed transmission of pea seedborne mosaic virus in pea is correlated with small size, abnormal shape, or seeds with cracked seed coats, but not with normal seeds.[112] Squash mosaic virus is transmitted at a higher percentage in lightweight, poorly filled, and deformed squash seeds compared to heavy, well-filled ones.[113] The separation of lettuce mosaic virus-infected lettuce seeds into heavy and lightweight portions by a vertical airstream concentrates the virus-containing seeds in the lightweight portions.[114] It is in contrast to reports that separation of lettuce seeds into light and heavy fractions by aspirations or gravity failed to increase or decrease the percentage of lettuce mosaic virus in any of the fractions.[21]

VII. SEED GERMINATION

Seed germination and emergence can favor seed transmission of pathogens by seedling growth rate. Slow growth may favor rapid infection or may limit infection. Cotyledons may carry the seed coat along with it or the seed coat may remain in the soil. In either case, depending upon the pathogen involved, it can effect infection of aerial parts. There are two patterns of seedling emergence.

A. Epigeal (or Epigeous)

The cotyledons and enclosed plumule (epicotyl) are carried up during hypocotyl growth, emerge from the soil, and become green and photosynthetic (see Figure 31A). Ultimately, cotyledons wither and drop. Epigeal germination occurs, for example, in seeds of castor bean, cucumber, French bean, lettuce, onion, peanut, and soybean. Epigeal germination favors infection of aerial parts by *Corynebacterium michiganense* pv. *michiganense* in tomato, *Pseudomonas syringae* pv. *phaseolicola* in beans, *Xanthomonas campestris* pv. *campestris* in cabbage, and *X. campestris* pv. *carotae* in carrot.[115]

In bottle gourd, the seed coat sometimes remains attached to "pegs" on seedlings after germination and serves as a source of primary inoculum for *Fusarium oxysporum* f. sp. *lagenarium*. The infection rate of seedlings with pegs accompanied by a seed coat is 14 to 18%, while for those having their seed coats carried by the cotyledons it is 2 to 3%. During the germination process, the fungus, which is latent prior to germination in the seed coat, multiples and penetrates the seedling from the lower side of the epidermis.[116]

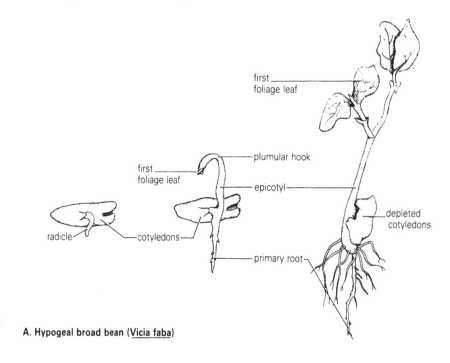

A. Hypogeal broad bean (Vicia faba)

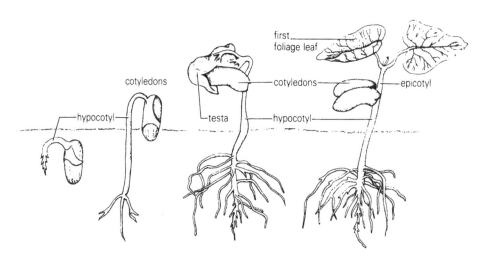

B. Epigeal French bean (Phaseolus vulgaris)

FIGURE 31.   Two types of germination. (A) Hypogeal germination of broad or horse bean (*Vicia faba*); (B) epigeal germination of bean (*Phaseolus vulgaris*). (After Bewley, J. D. and Black, M., *Physiology and Biochemistry of Seeds in Relation to Germination*, Part I, *Development, Germination and Growth*, Springer-Verlag, New York, 1978, 125.)

B. Hypogeal (or Hypogeous)

Cotyledons remain in the soil as the plumule (epicotyl) elongates and emerges (see Figure 31B). Hypogeal germination occurs, for example, in seeds of barley, broad-bean, maize, pea, and wheat. It favors root and stem infection. This type of germination limits seed transmission of bacteria that infect only aerial parts such as *C. fascians* in *Tropaeolum majus*,[117] *P. syringae* pv. *pisi* in pea,[118] *X. campestris* pv. *translucens* in cereals,[60] and *C. michiganense* pv. *nebraskense* in maize.[119]

Seedling infection, especially by the smut fungi, is influenced greatly by germination rate. The susceptibility of sorghum to *Sphacelotheca sorghi* is limited to the period between sowing and seedling emergence. The length of time of susceptibility depends upon the sowing date.[120]

VIII. SEED LEACHATES

Seeds produce sugars and amino acids during early stages of imbibition.[121] The chemical composition of leachates may interfere with spore germination.[122] Seed leachates of oil crops, cumin, and chilli pepper inhibit spore germination of many seedborne fungi.[123-125] Six sugars, twelve amino acids, and five organic acids are found in seed leachates of *Vigna radiata* and *V. mungo*. The interaction between the abiotic and biotic environment is mediated by the occurrence of seed leachates in relation to seedborne diseases.[126]

Seed leachates directly affect seedborne fungi by contributing to their nutritional status prior to penetration, or by inhibiting their saprophytic or pathogenic activity. Leachates from *V. radiata* retard the spore germination of *Alternaria alternata*, *Aspergillus flavus*, *Colletotrichum truncatum*, *Fusarium moniliforme*, *F. oxysporum*, *F. poae*, and *F. semitectum*. They stimulate spore germination of *Curvularia lunata*, *Drechslera rostrata*, *Phoma lingam*, etc.[127] Seed coat leachates of pigeon pea inhibit spore germination of *C. geniculata*, *D. hawaiiensis*, and *F. oxysporum*. The inhibitory action of seed coat leachates on fungal spore germination may be due to antifungal substances which may act as defensive agents against seed infection or seed transmission.[128 130]

IX. PRESENCE OF OTHER MICROFLORA

Seeds harbor a wide range of microflora and viruses and some of them affect seed transmission and establishment of infection in the field due to antagonistic action. The resistance of an oat cultivar from Brazil to *Drechslera victoriae* is due to antagonistic mycoflora such as *Chaetomium cochlioides* and *C. globosum* on the seed surface.[131] When infected seeds are planted in soil, these fungi produce a metabolite, cochliodinol.[132] There are indications that microorganisms antagonistic to *Pyrenophora avenae* in oats can inhibit seed transmission.[133] Bamberg et al.[134] demonstrated an interaction between *Tilletia caries* and *T. foetida,* both of which can infect alone but eliminate each other when in combination. Seeds inoculated with *T. foetida* and sown in soil infested with *T. caries* result in reduced seedling infection by *T. caries* compared to noninoculated seeds. In contrast, Berend[135] found no evidence of antagonism when seeds were inoculated with a mixture of the two fungi. *Corynebacterium michiganense* pv. *tritici* is unable to cause tundu disease in wheat in absence of the nematode *Anguina tritici.*[87] An antagonistic bacterium in sesame seeds inhibits the growth of *Pseudomonas syringae* pv. *sesami* in culture.[136] Cowpea stunt of cowpea is caused by a combination of cucumber mosaic virus and blackeye cowpea mosaic virus, which are synergistic. Both are seed transmitted and carried by the same aphid alone or in combination. The viruses can be transmitted from double-infected plants either separately or together, and cause single or double infections, respectively, in cowpeas.[137]

REFERENCES

1. Richardson, M. J., *An Annotated List of Seed Borne Diseases,* Commonwealth Mycological Institute, Kew, England, 1979, 320.
2. Bever, W. M., Embryo test not reliable for determining percentage of loose smut infection in wheat, *Ill. Res.,* 2, 18, 1960.
3. Gaskin, T. A. and Schafer, J. F., Some histological and genetic relationships of resistance of wheat to loose smut, *Phytopathology,* 52, 602, 1962.
4. Mantle, P. G., Further observations on an abnormal reaction of wheat to loose smut, *Trans. Br. Mycol. Soc.,* 44, 529, 1960.
5. Ohms, R. E. and Bever, W. M., Effect of *Ustilago tritici* infection in third internode elongation in resistant and susceptible winter wheat, *Phytopathology,* 44, 500, 1954.
6. Oort, A. J. F., Hypersensitivity of wheat to loose smut, *Tijdschr. Plantenziekten,* 50, 73, 1944.
7. Batts, C. C. V. and Jeater, A., The reaction of wheat varieties to loose smut as determined by embryo, seedling and adult plant tests, *Ann. Appl. Biol.,* 46, 23, 1958.
8. Hewett, P. D., Resistance to barley loose smut (*Ustilago nuda*) in the variety Emir, *Trans. Br. Mycol. Soc.,* 59, 330, 1972.
9. Mumford, D. L. E. and Rasmusson, D. C., Resistance of barley to *Ustilago nuda* after embryo infection, *Phytopathology,* 53, 125, 1963.
10. Ribeiro, V. do. M. A. M., Investigations into the pathogenicity of certain races of *Ustilago nuda, Trans. Br. Mycol. Soc.,* 46, 49, 1963.
11. Ling, L., The histology of infection of susceptible and resistant selfed lines of rye by the rye smut fungus, *Urocystis occulta, Phytopathology,* 30, 926, 1940.
12. Western, J. H., The biology of oat smuts. IV. The invasion of some susceptible and resistant oat varieties, including Markton, by selected biological species of smut (*Ustilago avenae* (Pers.) Jens. and *Ustilago kolleri* Wille), *Ann. Appl. Biol.,* 23, 245, 1936.
13. Ponirovskii, V. N., On the histology of the parasitism of *U. hordei* in barley shoots, *Tr. Khark. Skh. Inst.,* 38, 157, 1962.
14. Heald, F. D., The relation of spore load to the percent of stinking smut appearing in the crop, *Phytopathology,* 11, 269, 1921.
15. Woolman, H. M., Infection phenomena and host reactions caused by *Tilletia tritici* in susceptible and nonsusceptible varieties of wheat, *Phytopathology,* 20, 637, 1930.
16. Laurence, J. A. and Kennedy, B. W., Population changes of *Pseudomonas glycinea* on germinating soybean seeds, *Phytopathology,* 64, 1470, 1974.
17. Fajardo, T. G., Progress on experimental work with the transmission of bean mosaic, *Phytopathology,* 18, 155, 1928.
18. Smith, F. L. and Hewitt, W. B., Varietal susceptibility to common bean mosaic and transmission through seed, *Calif. Agric. Exp. Stn. Bull.,* 621, 18, 1938.
19. McNeal, F. H. and Afanasiev, M. M., Transmission of barley stripe mosaic through the seed in eleven varieties of spring wheat, *Plant Dis. Rep.,* 39, 460, 1955.
20. Hemmati, K. and McLean, D. L., Gamete-seed transmission of alfalfa mosaic virus and its effect on seed germination and yield in alfalfa plants, *Phytopathology,* 67, 576, 1977.
21. Grogan, R. G. and Bardin, R., Some aspects concerning the seed transmission of lettuce mosaic virus, *Phytopathology,* 40, 965, 1950.
22. Babovic, M. V., The transmission rate of alfalfa mosaic virus by lucerne seed, *Acta Biol. Yugosl.,* 13, 83, 1976.
23. Grogan, R. G. and Schnathorst, W. C., Tobacco ring spot virus — the cause of lettuce calico, *Plant Dis. Rep.,* 39, 803, 1955.
24. Grogan, R. G., Hall, D. H., and Kimble, K. A., Cucurbit mosaic viruses in California, *Phytopathology,* 49, 366, 1959.
25. Kennedy, B. W. and Cooper, R. L., Association of virus infection with mottling of soybean seed coats, *Phytopathology,* 57, 35, 1967.
26. Adams, D. B. and Kuhn, C. W., Seed transmission of peanut mottle virus, *Phytopathology,* 67, 1126, 1977.
27. Kendrick, J. B. and Gardner, M. W., Soybean mosaic seed transmission and effect on yield, *J. Agric. Res.,* 27, 91, 1924.
28. Athow, K. L. and Bancroft, J. B., Development and transmission of tobacco ringspot virus in soybean, *Phytopathology,* 49, 697, 1959.
29. Carroll, T. W. and Chapman, S. R., Variation in embryo infection and seed transmission of barley stripe mosaic virus within and between two cultivars of barley, *Phytopathology,* 60, 1079, 1970.
30. Timian, R. G., The range of symbiosis of barley and barley stripe mosaic virus, *Phytopathology,* 64, 342, 1974.

31. Jones, E. S., Influence of temperature, moisture and oxygen on spore germination of *Ustilago avenae*, *J. Agric. Res.*, 24, 577, 1923.
32. Reed, G. M. and Faris, J. A., Influence of environmental factors on the infectio.   sorghums and oats by smuts, *Am. J. Bot.*, 11, 518, 1924.
33. Kulkarni, G. S., Conditions influencing the distribution of grain smut (*Sphacelotheca sorghi*) of jowar (sorghum) in India, *Agric. J. India*, 17, 159, 1922.
34. Faris, J. A., Factors influencing infection of *Hordeum sativum* by *Ustilago hordei*, *Am. J. Bot.*, 11, 189, 1924.
35. Ling, L., Factors affecting infection in rye smut and subsequent development of the fungus in the host, *Phytopathology*, 31, 617, 1941.
36. Rabien, H., On the germination and infection conditions of *Tilletia tritici*, *Arb. Biol. Reichsanst. Land Forstwirtsch. Berlin Dahlem*, 14, 297, 1924.
37. Gibs, W., Modifications in susceptibility to bunt due to external conditions, *J. Landwirtsch.*, 72, 111, 1924.
38. Kendrick, E. L. and Purdy, L. H., Influence of environmental factors on the development of wheat bunt in the Pacific Northwest. III. Effect of temperature on time and establishment of infection by races of *Tilletia caries* and *T. foetida*, *Phytopathology*, 52, 621, 1962.
39. Colhoun, J., Taylor, G. S., and Tomlinson, R., Fusarium diseases of cereals. II. Infection of seedlings by *F. culmorum* and *F. avenaceum* in relation to environmental factors, *Trans. Br. Mycol. Soc.*, 51, 397, 1968.
40. Malalasekera, R. A. P. and Colhoun, J., Fusarium diseases of cereals. III. Water relations and infection of wheat seedlings by *Fusarium culmorum*, *Trans. Br. Mycol. Soc.*, 51, 711, 1968.
41. Millar, C. S. and Colhoun, J., Fusarium diseases of cereals. VI. Epidemiology of *Fusarium nivale* on wheat, *Trans. Br. Mycol. Soc.*, 52, 195, 1969.
42. Holmes, S. J. I. and Colhoun, J., Infection of wheat seedlings by *Septoria nodorum* in relation to environmental factors, *Trans. Br. Mycol. Soc.*, 57, 493, 1971.
43. Skoropad, W. P., Seed and seedling infection of barley by *Rhynchosporium secalis*, *Phytopathology*, 49, 623, 1959.
44. Leukel, R. W., Studies on bunt or stinking smut of wheat and its control, *U.S. Dep. Agric. Tech. Bull.*, 582, 47, 1937.
45. Purdy, L. H. and Kendrick, E. L., Influence of environmental factors on the development of wheat bunt in the Pacific Northwest. IV. Effect of soil temperature and soil moisture on infection by soil-borne spores, *Phytopathology*, 53, 416, 1963.
46. Schafer, J. F., Dickson, J. G., and Shands, H. L., Effect of temperature on covered smut expression in two barley varieties, *Phytopathology*, 52, 1161, 1962.
47. Leukel, R. W., Factors influencing infection of barley by loose smut, *Phytopathology*, 26, 630, 1936.
48. Purdy, L. H., Soil moisture and soil temperature, their influence on infection by the wheat flag smut fungus and control of the disease by three seed-treatment fungicides, *Phytopathology*, 56, 98, 1966.
49. Kavanagh, T., Temperature in relation to loose smut in barley and wheat, *Phytopathology*, 51, 189, 1961.
50. Dean, W. M., The effect of temperature on loose smut of wheat (*Ustilago nuda*), *Ann. Appl. Biol.*, 64, 75, 1969.
51. Hsi, C. H., Environment and sorghum kernel smut, *Phytopathology*, 48, 22, 1958.
52. Melchers, L. E. and Hansing, E. D., The influence of environmental conditions at planting time on sorghum kernel smut infection, *Am. J. Bot.*, 25, 17, 1938.
53. Joshi, L. M., Singh, D. V., and Srivastava, K. D., Meteorological conditions in relation to incidence of Karnal bunt of wheat in India, in *3rd Int. Symp. Plant Pathology*, Indian Agricultural Research Station, New Delhi, 1981, 11.
54. Agarwal, V. K., Singh, O. V., and Singh, A., A note on certification standard for the karnal bunt disease of wheat, *Seed Res.*, 1, 96, 1973.
55. Teviotdale, B. L. and Hall, D. H., Factors affecting inoculum development and seed transmission of *Helminthosporium gramineum*, *Phytopathology*, 66, 295, 1976.
56. Lehman, S. G., Systemic infection of soybean by *Peronospora manshurica* as affected by temperature, *J. Elisha Mitchell Sci. Soc.*, 69, 83, 1953.
57. Kommedahl, T. and Lang, D. S., Temperature effects on seedling wilt from corn kernels infected with *Helminthosporium maydis*, *Phytopathology*, 62, 770, 1972.
58. Calvert, O. H. and Thomas, C. A., Some factors affecting seed transmission of safflower rust, *Phytopathology*, 44, 609, 1954.
59. Shetty, H. S., Mathur, S. B., and Neergaard, P., Occurrence of *Sclerospora graminicola* (Sacc.) Schroet. inoculum in pearl millet (*Pennisetum typhoides* (Burm.) Stapf and Hubb.) seeds and its transmission, in *3rd Int. Congr. Plant Pathology*, P. Parey, Berlin, 1978, 120.
60. Wallin, J. R., Seed and seedling infection of barley, bromegrass, and wheat by *Xanthomonas translucens* var. *cerealis*, *Phytopathology*, 36, 446, 1946.

61. Crowley, N. C., Studies on the seed transmissions of plant virus diseases, *Aust. J. Biol. Sci.*, 10, 449, 1957.
62. Singh, G. P., Arny, D. C., and Pound, G. S., Studies on the stripe mosaic of barley, inᴜ of temperature and age of host on disease development and seed infection, *Phytopatholoᵹ* 1960.
63. Agarwal, V. K., Nene, Y. L., and Beniwal, S. P. S., Detection of bean common mosaic virus in urdbean (*Phaseolus mungo*) seeds, *Seed Sci. Technol.*, 5, 619, 1977.
64. Hanada, K. and Harrison, B. D., Effects of virus genotype and temperature on seed transmission of nepoviruses, *Ann. Appl. Biol.*, 85, 79, 1977.
65. Daft, G. C. and Leben, C., Bacterial blight of soybeans: epidemiology of blight outbreaks, *Phyto-pathology*, 62, 57, 1972.
66. Guthrie, J. W., Factors influencing halo blight transmission from externally contaminated *Phaseolus vulgaris* seed, *Phytopathology*, 60, 371, 1970.
67. Grogan, R. G., Lucas, L. T., and Kimble, K. A., Angular leaf spot of cucumber in California, *Plant Dis. Rep.*, 55, 3, 1971.
68. Rodenhiser, H. A. and Taylor, J. W., Studies on environmental factors affecting infection and the development of bunt in wheat, *Phytopathology*, 30, 20, 1940.
69. Rodenhiser, H. A. and Taylor, J. W., The effect of photoperiodism on the development of bunt in two spring wheats, *Phytopathology*, 33, 240, 1943.
70. Polyakov, I. M. and Vladimirskaya, M. E., The role of light conditions in the resistance of cabbage to false powdery mildew, *Tr. Vses. Inst. Zashch. Rast.*, 21, 18, 1964.
71. Glaeser, G., The extent of field attack by wheat bunt (*T. caries*) in relation to seed infestation, *Pflasch Ber.*, 26, 33, 1961.
72. Malalasekera, R. A. P. and Colhoun, J., Fusarium diseases of cereals. V. A technique for the examination of wheat seed infected with *Fusarium culmorum*, *Trans. Br. Mycol. Soc.*, 52, 187, 1969.
73. Aulakh, K. S., Mathur, S. B., and Neergaard, P., Comparison of seed-borne infection of *Drechslera oryzae* as recorded on blotter and in soil, *Seed Sci. Technol.*, 2, 385, 1974.
74. Wallen, V. R., Field evaluation and the importance of the *Ascochyta* complex on peas, *Can. J. Plant Sci.*, 45, 27, 1965.
75. Moffett, M. L., Wood, B. A., and Hayward, A. C., Seed and soil: sources of inoculum for the colonisation of the foliage of Solanaceous hosts by *Pseudomonas solanacearum*, *Ann. Appl. Biol.*, 98, 403, 1981.
76. Popp, W., A new approach to the embryo test for predicting loose smut of wheat in adult plants, *Phytopathology*, 49, 75, 1959.
77. Khanzada, A. K., Rennie, W. J., Mathur, S. B., and Neergaard, P., Evaluation of two routine embryo test procedures for assessing the incidence of loose smut infection in seed samples of wheat (*Triticum aestivum*), *Seed Sci. Technol.*, 8, 363, 1980.
78. Maude, R. B. and Humpherson-Jones, F. M., Studies on the seed-borne phases of dark leaf spot (*Alternaria brassicicola*) and grey leaf spot (*Alternaria brassicae*) of Brassicas, *Ann. Appl. Biol.*, 95, 311, 1980.
79. Zade, A., Recent investigation on the life-history and control of loose smut of oats [*Ustilago avenae* (Pers.) Jens.], *Angew. Bot.*, 6, 113, 1924.
80. Emdal, P. S. and Foldo, N. E., Seed borne inoculum of *Uromyces betae*, *Seed Sci. Technol.*, 7, 93, 1979.
81. Anderson, C. W., Seed transmission of three viruses in cowpea, *Phytopathology*, 47, 515, 1957.
82. Cheo, P. C., Effect of seed maturation on inhibition of southern bean mosaic virus in bean, *Phyto-pathology*, 45, 17, 1955.
83. Shepherd, R. J. and Fulton, R. W., Identity of a seed-borne virus of cowpea, *Phytopathology*, 52, 489, 1962.
84. Nelson, M. R. and Knuhtsen, H. K., Squash mosaic virus variability: epidemiological consequences of differences in seed transmission frequency between strains, *Phytopathology*, 63, 918, 1973.
85. Nelson, M. R. and Knuhtsen, H. K., Relation of seed transmission to the epidemiology of squash mosaic virus strains, *Phytopathology*, 59, 1042, 1969.
86. Ghanekar, A. M. and Schwenk, F. W., Seed transmission and distribution of tobacco streak virus in six cultivars of soybean, *Phytopathology*, 64, 112, 1974.
87. Midha, S. K. and Swarup, G., Factors affecting development of ear-cockle and tundu diseases of wheat, *Indian J. Nematol.*, 2, 97, 1972.
88. Baker, C. J., Morphology of seedling infection by *Leptosphaeria nodorum*, *Trans. Br. Mycol. Soc.*, 56, 306, 1971.
89. Rodenhiser, H. A. and Taylor, J. W., Effects of soil type, soil sterilization and soil reaction on bunt infection at different incubation temperatures, *Phytopathology*, 30, 400, 1940.
90. Doling, D. A., The influence of seedling competition on the amount of loose smut (*Ustilago nuda* (Jens.) Rostr.) appearing in barley crops, *Ann. Appl. Biol.*, 54, 91, 1964.

91. Milan, A., Sul "carbone volante" del grano in rapporto all'accestimento delle piante, *Nuovo G. Bot. Ital.*, 46, 149, 1939.
92. Jones, G. H. and Seif-El-Nasr, A. El G., The influence of sowing depth and moisture on smut diseases and the prospects of a new method of control, *Ann. Appl. Biol.*, 27, 35, 1940.
93. Swinburne, T. R., Infection by *Tilletia caries* (DC.) Tul., the causal organism of bunt, *Trans. Br. Mycol. Soc.*, 46, 145, 1963.
94. Tiemann, A., Untersuchungen über die Empfanglichkeit des Sommerweizens für *Ustilago tritici* und den Einfluss der äusseren Bedingungen dieser Krankheit, *Kuhn-Arch.*, 9, 405, 1925.
95. Pedersen, P. N., Investigations on the influence of growth conditions on the attacks of loose smut of barley, *Acad. Scand.*, 15, 245, 1965.
96. Meiners, J. P., Kendrick, E.L., and Holton, C. S., Depth of seeding as a factor in the incidence of dwarf bunt and its possible relationship to spore germination on or near the soil surface, *Plant Dis. Rep.*, 40, 242, 1956.
97. Leukel, R. W., Investigations on the nematode disease of cereals caused by *Tylenchus tritici*, *J. Agric. Res.*, 27, 925, 1924.
98. Hewett, P. D., Loose smut in winter barley: comparisons between embryo infection and the production of diseased ears in the field, *J. Natl. Inst. Agric. Bot.*, 15, 231, 1980.
99. Slack, S. A., Shepherd, R. J., and Hall, D. H., Spread of seed borne barley stripe mosaic virus and effects of the virus on barley in California, *Phytopathology*, 65, 1218, 1975.
100. Gassner, G. and Kirchhoff, H., Zur Frage der Beeinflussung des Flugbrandbefalls durch Umweltfaktoren und chemische Beizmittel, *Phytopathol. Z.*, 7, 487, 1934.
101. Olsson, L., The influence of certain factors on the occurrence of seedling injuring fungi in the resulting crop of cereal seed, *Seed Sci. Technol.*, 7, 235, 1979.
102. Ooka, J. J. and Kommedahl, T., Kernel infected with *Fusarium moniliforme* in corn cultivars with opaque-2-endosperm or male sterile cytoplasm, *Plant Dis. Rep.*, 61, 162, 1977.
103. Templeton, G. E., Johnson, T. H., and Henry, S. E., Kernel smut of rice, *Ark. Farm Res.*, 9, 10, 1960.
104. Krull, C. F., Robayo, G., Valbuena, L. A., Luis, A., Rico, G., Castibalco, L. E., and Bravo, L. E., Influence of seed size on the incidence of loose smut in Funza barley, *Plant Dis. Rep.*, 50, 101, 1966.
105. Kuznetsova, A. F., The effect of seed size on infection of barley by loose smut, *Ref. Zh. Rastenievod.*, 855, 1971.
106. Lavery, P., The relationship between seed size and the incidence of loose smut in three winter barleys, *Proc. Indian Acad. Sci.*, 74, 155, 1965.
107. Taylor, J. W., Effect of the continuous selection of large and small wheat seed on yield, bushel weight, varietal purity, and loose smut infection, *J. Am. Soc. Agron.*, 20, 856, 1928.
108. McFadden, A. D., Kaufmann, M. L., Russell, R. C., and Tyner, L. E., Association between seed size and the incidence of loose smut in barley, *Can. J. Plant Sci.*, 40, 611, 1960.
109. Agarwal, V. K., Seed-borne fungi and viruses of some important crops, *Res. Bull.* No. 108, Experimental Station, G. B. Pant University of Agricultural Technology, Pantnagar, India, 1981, 144.
110. Paguio, O. R. and Kuhn, C. W., Incidence and sources of inoculum of peanut mottle virus and its effect on peanut, *Phytopathology*, 64, 60, 1974.
111. Troutman, J. L., Bailey, W. K., and Thomas, C. A., Seed transmission of peanut stunt virus, *Phytopathology*, 57, 1280, 1967.
112. Stevenson, W. R. and Hagedorn, D. J., Further studies on seed transmission of pea seed-borne mosaic virus in *Pisum sativum*, *Plant Dis. Rep.*, 57, 248, 1973.
113. Middleton, J. T., Seed transmission of squash mosaic virus, *Phytopathology*, 34, 405, 1944.
114. Ryder, E. J. and Johnson, A. S., A method for indexing lettuce seeds for seed borne lettuce mosaic virus by airstream separation of light from heavy seeds, *Plant Dis. Rep.*, 58, 1037, 1974.
115. Schuster, M. L. and Coyne, D. P., Survival mechanism of phytopathogenic bacteria, *Annu. Rev. Phytopathol.*, 12, 199, 1974.
116. Kuniyasu, K., Seed transmission of Fusarium wilt of bottle gourd *Lagenaria siceraria*, used as root stock of watermelon. III. Course of seedling infection by the seed borne pathogen, *Fusarium oxysporum* f. sp. *lagenarium* Matuo and Yamamoto, *Ann. Phytopathol. Soc. Jpn.*, 43, 270, 1977.
117. Baker, K. F., Bacterial fasciation disease of ornamental plants in California, *Plant Dis. Rep.*, 34, 121, 1950.
118. Skoric, V., Bacterial blight of peas: overwintering, dissemination, and pathological histology, *Phytopathology*, 17, 611, 1927.
119. Schuster, M. L., Hoff, B., Mandel, M., and Lazar, I., Leaf freckles and wilt, a new corn disease, *Proc. Annu. Corn Sorghum Res. Conf. (Chicago)*, 27, 176, 1973.

120. El-Helaly, A. F. and Ibrahim, I. A., Host parasite relationship of *Sphacelotheca sorghi* on sorghum, *Phytopathology*, 47, 620, 1957.

121. Larson, L. A. and Beenvers, H., Amino acid metabolism in young pea seedlings, *Plant Physiol.*, 40, 424, 1966.

122. Kandaswamy, D., Kesavan, R., Ramaswamy, K., and Prasad, N. N., Occurrence of microbial inhibitors in the exudates of certain leguminous seeds, *Indian J. Microbiol.*, 14, 25, 1974.

123. Chaturvedi, S. N., Muralia, R. N., and Sirdhana, B. S., The influence of cumin seed exudates on fungal spore germination, *Plant Soil*, 40, 49, 1974.

124. Dhawale, S. D. and Kodmelvar, R. V., Studies on mycoflora of chilli seed, *Seed Res.*, 6, 23, 1978.

125. Mishra, R. R. and Kanaujia, R. S., Studies on certain aspects of seed-borne fungi, *Indian Phytopathol.*, 26, 284, 1965.

126. Takayanki, K. and Murakame, K., Rapid germinability test with exudates from seed, *Proc. Int. Seed Test. Assoc.*, 34, 243, 1968.

127. Saxena, R. M. and Gupta, J. S., Effect of seed leachates on spore germination of seed-borne fungi on *Vigna radiata*, *Indian Phytopathol.*, 35, 236, 1982.

128. Charya, M. A. S. and Reddy, S. M., Effect of *Cajanus cajan* seed coat leachates on germination of some seedborne fungi, *Indian Phytopathol.*, 33, 112, 1980.

129. Ark, P. A. and Thompson, J. P., Antibiotic properties of the seeds of wheat and barley, *Plant Dis. Rep.*, 42, 959, 1958.

130. Srivastava, V. B. and Mishra, R. R., Fungal inhibitory agents in seed coats, *Phytopathol. Mediterr.*, 10, 127, 1971.

131. Tveit, M. and Moore, M. B., Isolates of *Chaetomium* that protect oats from *Helminthosporium victoriae*, *Phytopathology*, 44, 686, 1954.

132. Brewer, D., Jeram, W. A., Meiler, D., and Taylor, A., The toxicity of cochliodinol, an antibiotic metabolite of *Chaetomium* spp., *Can. J. Microbiol.*, 16, 433, 1970.

133. Old, K. M., Mercury tolerant *Pyrenophora avenae* in seed oats, *Trans. Br. Mycol. Soc.*, 51, 525, 1968.

134. Bamberg, R. H., Holton, C. S., Rodenhiser, H. A., and Woodward, R. W., Wheat dwarf bunt depressed by common bunt, *Phytopathology*, 37, 556, 1947.

135. Berend, I., The occurrence of bunt fungi in wheat inoculated by *Tilletia caries* and *T. foetida*, *Acta Phytopathol. Acad. Sci. Hung.*, 8, 365, 1973.

136. Vajavat, R. M. and Chakravarti, B. P., Survival of *Pseudomonas sesami* and effect of an antagonistic bacterium isolated from seeds on the control of the disease in seed, *Indian Phytopathol.*, 31, 286, 1978.

137. Pio-Ribeiro, G., Wyatt, S. D., and Kuhn, C. W., Cowpea stunt: a disease caused by a synergistic interaction of two viruses, *Phytopathology*, 68, 1260, 1978.

Chapter 10

EPIPHYTOLOGY OF SEEDBORNE DISEASES

I. INTRODUCTION

Epiphytology is the study of the development and spread of disease and of the factors affecting these processes,[1] and deals with the effects of the biotic (host and pathogen) and abiotic environment. Van der Plank[2] defined the study of epiphytotics as the science of disease in populations involving the persistence and spread of inoculum and environmental factors affecting disease incidence on particular plant populations. Epiphytological studies are important in disease forecasting and disease control. Van der Plank[2] applied mathematical models to epiphytological studies of plant diseases. Plant disease epiphytotic refers to the development and rapid spread of a disease on a particular kind of crop cultivated over a large area.[3] Epiphytotics develop when a susceptible cultivar is planted over a large area in the presence of a virulent pathogen coupled with a favorable environment. The development and decline of epiphytotics is a balance between inoculum potential and disease potential:

$$\text{Disease severity} \quad = \quad \underset{\text{(inoculum density} \times \text{capacity)}}{\text{Inoculum potential}} \quad \times \quad \underset{\text{(proneness} \times \text{susceptibility)}}{\text{Disease potential}}$$

Inoculum potential is the number of infective propagules (inoculum density) and their pathogenic capacity. Disease potential is host susceptibility, which may be influenced by an unfavorable environment, nutritional imbalances, and/or a susceptible growth stage. Host susceptibility is genetically controlled and development of an epiphytotic depends on the number of infective propagules, their pathogenic capacity, host susceptibility, and the effect of environment on the pathogen virulence and host proneness. These same factors affect the epiphytology of seedborne pathogens.

The rate of seed infection to seed transmission or plant infection and subsequent establishment in the field and further spread of inoculum depends on the host, pathogen(s), environment, transmitting agents, and their interactions. Yield losses may be high even with a low-percent seed infection by certain pathogens. For example, as few as 2 Brassica seeds per 10,000 infected with *Xanthomonas campestris* pv. *campestris*,[4] less than 1% *X. campestris* pv. *vesicatoria*-infected tomato seeds,[5] or 0.02% *Pseudomonas syringae* pv. *phaseolicola*-infected seed in bean[6] can produce disease epiphytotics caused by these pathogens under suitable environmental conditions. The incidence of cowpea blight (*X. campestris* pv. *vignicola*) may be 62% from an initial inoculum of 1% infected seeds.[7] If lettuce seed lots carry over 0.5% lettuce mosaic virus, significant yield losses may result if aphids are prevalent.[8]

Environmental conditions greatly influence the epiphytology of seedborne pathogens. During the summer in India (relative humidity 20 to 80%, and 25 to 34°C) the incidence of cowpea blight (*X. campestrix* pv. *vignicola*) reaches up to 62% from an initial inoculum load of 1% infected seed but the incidence of greengram leaf spot (*X. campestris* pv. *phaseoli*) is only 0, 3, and 32% from 1, 10 and 100% initial seed infestation. However, during the rainy season (relative humidity 50 to 95%, 24 to 30°C), both diseases become severe.[7] As few as 2 infected seeds per 10,000 can cause an epiphytotic of blackrot (*X. campestris* pv. *campestris*) in cabbage. However, the environmental conditions influence the spread of primary inoculum. In 1976, seed infestations of 0.12, 0.06, and 0.02% resulted in epiphytotics. In 1977, epiphytotics resulted from infestations of 0.05% but not 0.01%.[4] A population of 10^3 to 10^4 *X. campestris* pv.

phaseoli per bean seed is required for production of infected plants under field conditions.[9]

The threshold level for pathogenicity of any seedborne inoculum is not constant since it is affected by inoculum level, inoculum location, seed- and soilborne microflora, soil temperature, moisture, pH, environmental factors, insect vectors, wind velocity, etc. Epiphytotics vary with different pathogens. For example, *Ascochyta pinodella* and *Mycosphaerella pinodes* cause greater reductions in yield than *A. pisi*. *A. pisi* requires a higher incidence of seed infection than *A. pinodella* and *M. pinodes* to lower yields.[10] The ratio of seed infection to plant infection varies depending upon cultivar; the rate of seed infection to tomato plant infection of *Fusarium oxysporum* f. sp. *lycopersici* is 1:0.75, 1:0.70, and 1:0.36 for cultivars Monalbo, Eclaireur, and Exhibition, respectively.[11]

II. CLASSIFICATION OF SEEDBORNE DISEASES BASED ON EPIPHYTOLOGY

The seedborne disease can be classified into either monocyclic (simple-interest) diseases or polycyclic (compound-interest) disease based on their epiphytology.[2,12]

A. Monocyclic or Simple-Interest Diseases

These are diseases in which the seedborne pathogen is closely associated with the host after infection until symptom production, and reinfection of the host does not occur (Figure 32). However, inoculum transfer may result in seed infection. Epiphytotics due to such infections occur only when seeds are heavily contaminated or infected. Typical examples are barley and wheat loose smut and wheat bunts in which plants show symptoms only at ear formation. In Sonora-64, a susceptible genotype of wheat, a direct correlation was found between embryo and seedlings having more than 50% of the tissue invaded with hyphae and field expression of the disease.[13]

A system is being developed in India for predicting loose smut of wheat utilizing the seed infection and transmission ratio. Seed infection of wheat loose smut is determined by an embryo count and seed transmission ratio on tiller infection in the field. The data are fitted into the equation $Y = EK$ for calculating the predicted value of loose smut incidence, where Y = predicted value of loose smut incidence, E = percent loose smut infection in the seed lot, and K = the constant for seed transmission of loose smut in the cultivar used based on the previous 3 years.[14]

In the U.K., leaf stripe of barley (*Drechslera graminea*) does not reinfect leaves and is considered as a simple-interest disease. The initial seedborne inoculum is the total inoculum available each year.[15] The disease cycle of *Gloeotinia temulenta*, blind seed disease, and *Claviceps purpurea*, ergot, which invade only the inflorescence also are regarded as simple-interest diseases by Hewett,[15] but conidia produced by both fungi on infected plants may result in secondary spread. These diseases should be considered as compound-interest diseases.

It is easier to control simple-interest diseases because the use of clean seed can be achieved using seed certification or seed treatment with systemic or nonsystemic fungicides.

B. Compound-Interest or Polycyclic Diseases

These are diseases in which seedborne pathogens after infection produce inoculum on the host capable of being carried to other host plants which may result in disease (Figure 32). The seedborne inoculum gives rise to infected seedlings scattered throughout the field. The inoculum multiplies repeatedly and spreads throughout the growing period of the host when conditions are favorable.[2] The inoculum may be carried by

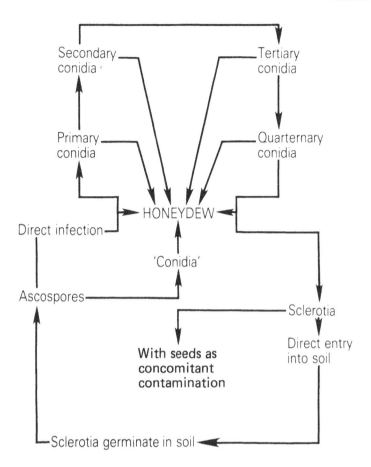

FIGURE 32. Life cycle of *Claviceps fusiformis* (ergot of pearlmillet (*Pennisetum typhoides*). (After Prakash, H. S., Shetty, H. S., and Safeeulla, K. M., *Trans. Br. Mycol. Soc.*, 81, 65, 1983.)

rain-splash, wind, vectors, and other means. Epiphytotics of compound-interest diseases are influenced by the environment. Under favorable conditions even a low initial seedborne inoculum can result in an epiphytotic. Many seedborne pathogens belong to this group. Some examples are given.

Bean seeds infected with *A. fabae* produce seedlings with leaf lesions, inoculum from which spreads for up to 10 m in an average season and usually infects the new seed crop. Seed lots with 1% infected seeds are suitable for crop production, but little or no *A. fabae* is tolerated in seeds intended for multiplication. Infection in British-grown commercial seeds is reduced by seed selection. Standards adopted in the field bean seed scheme eliminate *A. fabae*. Theoretically, 1% seed infection gives one diseased seedling per 40 m² (18 cm between rows).[16] *A. fabae* in *Vicia faba* increases from initial seed infection levels of 8.4% to 20% at harvest.[17]

Even if the rate of seed transmission is as low as 0.01%, as in the case of downy mildew of pearlmillet (*Sclerospora graminicola*), there could be 30 infected plants per acre. These plants may produce sporangial inoculum sufficient for secondary spread and cause an epiphytotic. In India, the seed rate for pearlmillet is 3 kg/acre (1000 grains in 10 g) which will give about 300,000 plants. A plant infected by *S. graminicola* can produce 35,000 sporangia per cm² of the infected leaf in 1 hr during night.[18]

Bacteria have a short generation time, thus even a low inoculum level can result in an epiphytotic.[19] For example, *P. syringae* pv. *phaseolicola* (halo blight of beans) is seedborne, and infected seeds may give rise to infected seedlings which serve as a source of primary inoculum (Figure 14). Rain-splash and wind help further disease spread. In the U.S., secondary infections are found up to 26 m away from primary infection sites.[5] Taylor[20] reported that only one of ten infected seeds produce infected seedlings. Rapid disease spread results in severe epiphytotics. Taylor et al.[21] used a Van der Plank[2] equation as an epiphytotic model for halo blight in bean:

$$r = \frac{2.3}{t_2 - t_1} \left(\log 10 \frac{X_2}{1 - x_2} - \log 10 \frac{X_1}{1 - x_1} \right)$$

where x_1 = the initial inoculum (or disease) at time t_1, x_2 = the inoculum (or disease) present at time t_2, and r = infection rate expressed as per unit per day (where $t_2 - t_1$ is measured in days). They proposed this equation as an appropriate one for a disease of the compound-interest type. The values of r will vary with climatic differences; from 0.17 to 0.22 and 0.10 to 0.16 in Wisconsin and the U.K., respectively.[6,22,23] Under British conditions it was predicted that 0.025% seed infection would give rise to 0.0025% primary infection and that this would produce 4% infection in the mature crop with an r-value of 0.15.[21] The model has been used to determine tolerance levels for seed infection and to compare the effectiveness of foliar sprays and seed treatments. A reduction in primary inoculum derived from infected seed by disease either by exclusion through seed testing or seed treatment gave effective disease control.[21] Seed treatment reduces initial inoculum (x_1) and foliar sprays affect the rate of disease increase (r). Effective control of halo blight is achieved by seed treatment with antibiotics streptomycin or kasugamycin, which reduce seed infection by 98%,[23] or with the foliar sprays with streptomycin sulfate or copper oxychloride.[22,23]

Five diseased seedlings per 10,000 crucifer seeds with *X. campestris* pv. *campestris* can result in a high incidence of black rot, but a single diseased seedling will not. Laboratory seed assays capable of detecting 1 infected seed per 10,000 can predict field severity of black rot. Thus, it is suggested that a tolerance of 1 infected seed per 10,000 be accepted for direct seeding of cabbage for head production, but a zero tolerance be accepted for seedbed production.[24,25] Infecting cabbage seeds with *X. campestris* pv. *campestris* up to 1.0% results in a high disease incidence in the direction of the prevailing wind.[26] The severity of black rot infection in brussels sprouts in central New York State is related inversely to the distance from infected transplants.[27]

Twelve bean seeds per acre infected with *P. syringae* pv. *phaseolicola* can cause a severe epiphytotic of halo blight. Diseased plants can be found 20 to 25 m from the inoculum source at harvest. Rain-splash dispersal disseminates the bacterium under Wisconsin conditions.[6]

Seed transmission is the primary source of inoculum and the most important factor in the epiphytology of certain plant viruses such as BSMV in barley and wheat, BCMV in bean, broadbean stain virus in broadbean, eggplant mosaic virus in eggplant, lettuce mosaic virus in lettuce, peanut mottle virus in peanut, squash mosaic virus in cantaloupe, TMV in tomato, and nepoviruses. The spread of SMV from soybean seedlings infected from seed depends on the percent of infected seedlings, timing, numbers and species composition of transient alate aphids, and wind and other environmental factors. The pattern of spread decreases as the distance from infected seedling increases. A spread up to 50 m was noted from an initial inoculum source.[28] In 1959—60, lettuce crops from seed <0.1% infected with lettuce mosaic virus contained 0 to 4.5% infected plants, whereas crops from 2.5 to 5.3% infected seeds contained 25 to 96% infected plants.[29]

A unique study of the annual occurrence of a seedborne fungus was made by Shortt et al.[30] In the 3-year study of seed decay of soybean caused by *Phomopsis* spp. in Illinois, disease incidence was highest in 1977, lowest in 1976, and intermediate in 1975.

A low positive correlation was found between temperature and disease incidence, but no consistent continuum of disease from north to south within the state was apparent. The highest incidence of Phomopsis seed decay occurred along major waterways in the wet years of 1975 and 1977. A high positive correlation was found between disease incidence and rainfall during pod fill, indicating that moisture, rather than temperature or geographic area, is the dominant environmental factor in disease development. Maturity dates of cultivars interacted with changing weather conditions to affect disease incidence.

REFERENCES

1. Zakods, J. C. and Schein, R. D., *Epidemiology and Plant Disease Management,* Oxford University Press, London, 1979.
2. Van der Plank, J. E., *Plant Disease: Epidemics and Control,* Acaedmic Press, New York, 1963, 349.
3. Agrios, G. N., *Plant Pathology,* Academic Press, New York, 1978, 703.
4. Schaad, N. W., Sitterly, W. R., and Humaydan, H., Relationship of levels of seed-borne *Xanthomonas campestris* as determined by laboratory assays to the development of black rot of cabbage, 3rd Int. Congr. Plant Pathology, Munich, 1978, 66.
5. Cox, R. S., The role of bacterial spot in tomato production in South Florida, *Plant Dis. Rep.,* 50, 699, 1966.
6. Walker, J. C. and Patel, P. N., Splash dispersal and wind as factors in epidemiology of halo blight of bean, *Phytopathology,* 54, 140, 1964.
7. Shekhawat, G. S. and Patel, P. N., Seed transmission and spread of bacterial blight of cowpea and leaf spot of greengram in summer and monsoon seasons, *Plant Dis. Rep.,* 61, 390, 1977.
8. Broadbent, L., Tinsley, T. W., Buddin, W., and Roberts, E. T., The spread of lettuce mosaic in the field, *Ann. Appl. Biol.,* 38, 689, 1951.
9. Weller, D. M. and Saettler, A. W., Evaluation of seed-borne *Xanthomonas phaseoli* and *X. phaseoli* var. *fuscans* as primary inocula in bean blights, *Phytopathology,* 70, 148, 1980.
10. Wallen, V. R., Field evaluation and the importance of the *Ascochyta* complex on peas, *Can. J. Plant Sci.,* 45, 27, 1965.
11. Besri, M., Phases of the transmission of *Fusarium oxysporum* f. sp. *lycopersici* and *Verticillium dahliae* by seeds of some tomato varieties, *Phytopathol. Z.,* 93, 148, 1978.
12. Hewett, P. D., Epidemiology of seed-borne disease, in *Plant Disease Epidemiology,* Scott, P. R. and Bainbridge, A., Eds., Blackwell Scientific, Oxford, 1978, 167.
13. Rewal, H. S. and Jhooty, J. S., Correlation between embryo, seedling and field infection of loose smut of wheat, *Indian Phytopathol.,* 35, 571, 1982.
14. Verma, H. S., Singh, A., and Agarwal, V. K., A system for prediction of loose smut incidence in wheat, 3rd Int. Symp. Plant Pathology, New Delhi, 1981, 5.
15. Hewett, P. D., Disease testing in a seed improvement programme, in *Seed Pathology — Problems and Progress,* Yorinori, J. T., Sinclair, J. B., Mehta, Y. R., and Mohan, S. K., Eds., Fundação Instituto Agronômico do Paraná — IAPAR, Londrina, Brazil, 1979, 72.
16. Hewett, P. D., The field behaviour of seed-borne *Ascochyta fabae* and disease control in field beans, *Ann. Appl. Biol.,* 74, 287, 1973.
17. Gaunt, R. E. and Liew, R. S., Control strategies of *Ascochyta fabae* in New Zealand field and broadbean crops, *Seed Sci. Technol.,* 9, 707, 1981.
18. Safeeulla, K. M., *Biology and Control of the Downy Mildews of Pearlmillet, Sorghum, and Finger Millet,* Downy Mildew Research Laboratory, Mysore University, Mysore, India, 1976, 304.
19. Schuster, M. L. and Coyne, D. P., Survival mechanism of phytopathogenic bacteria, *Annu. Rev. Phytopathol.,* 12, 199, 1974.
20. Taylor, J. D., The quantitative estimation of the infection of bean seed with *Pseudomonas phaseolicola* (Burkh.) Dowson., *Ann. Appl. Biol.,* 66, 29, 1970.
21. Taylor, J. D., Dudley, C. L., and Presly, L., Studies of halo-blight seed infection and disease transmission in dwarf beans, *Ann. Appl. Biol.,* 93, 267, 1979.
22. Taylor, J. D., Field studies on halo-blight of beans (*Pseudomonas phaseolicola*) and its control by foliar sprays, *Ann. Appl. Biol.,* 70, 191, 1972.
23. Taylor, J. D. and Dudley, C. L., Seed treatment for the control of halo blight of beans (*Pseudomonas phaseolicola*), *Ann. Appl. Biol.,* 85, 223, 1977.

24. Schaad, N. W. and Kendrick, R., A qualitative method of detecting *Xanthomonas campestris* in crucifer seed, *Phytopathology,* 65, 1034, 1975.
25. Schaad, N. W., Sitterly, W. R., and Humaydan, H., Relationship of incidence of seed-borne *Xanthomonas campestris* to black rot of crucifers, *Plant Dis.,* 64, 91, 1980.
26. Cytrynomicz, L. E. and Fieldhouse, D. J., Epidemiology of *Xanthomonas campestris* on field grown cabbage, *Phytopathology,* 70, 688, 1980.
27. Hunter, J. E., Abawi, G. S., and Becker, R. F., Observations on the source and spread of *Xanthomonas campestris* in an epidemic of black rot in New York, *Plant Dis. Rep.,* 59, 384, 1975.
28. Irwin, M. E. and Goodman, R. M., Ecology and control of soybean mosaic virus, in *Plant Disease and Vectors,* Maramorosch, K. and Harris, K. F., Eds., Academic Press, New York, 1981, 181.
29. Tomlinson, J. A., Control of lettuce mosaic by the use of healthy seed, *Plant Pathol.,* 11, 61, 1962.
30. Shortt, B. J., Grybauskas, A. P., Tenne, F. D., and Sinclair, J. B., Epidemiology of Phomopsis seed decay of soybean in Illinois, *Plant Dis.,* 65, 62, 1981.

Chapter 11

NONPARASITIC SEED DISORDERS

Nontransmissible seed disorders are nonpathogenic, noninfectious, and may or may not lead to deleterious effects on seeds or seedlings. Nontransmissible disorders are due to a number of factors.

I. GENETIC EFFECTS

The genetic background of a host genotype can give rise to seed abnormalities. For example, in flax, seed coat splitting in yellow-seeded cultivars is common.[1] Eggplant, pepper, and tomato seed breaks commonly occur in the seed coat, resulting in a chamber between the seed coat and endosperm, which provides an entry point for pathogens.[2,3] Defective fertilization in barley and wheat can result in production of embryoless seeds.[4,5] The fruit pox and gold fleck diseases of tomato are inherited and can be transmitted through seeds for four generations. Fruit pox symptoms appear as incipient lesions on immature, green fruit, which later rupture and become necrotic before fruits turn red. Gold fleck symptoms appear as small lesions on immature, green fruit, which turn golden yellow on ripe fruit. Fruit pox is controlled by a recessive gene and gold fleck by a dominant gene. It is possible to eliminate fruit pox and gold fleck within two generations by rouging affected plants in segregating populations. A cultivar, Florida MH-1, free from fruit pox and gold fleck was released using this approach.[6]

Stevenson et al.[7] reported a seed-transmitted genetic disease in Chico III tomato known as corky stunt, which is expressed as shortened internodes, proliferated axillary buds, malformed petioles, roughened lower petiole surface, and malformed and corked fruit. Corky lesions blemish the fruit surface. The optimum temperature for symptom expression is 22.5°C. The disease is controlled by a single recessive gene.

II. MECHANICAL INJURIES

Operations such as harvesting, threshing, processing, and postharvest handling may result in mechanical injury to seeds (Figure 33). Injuries range from cracking or splitting of the seed coat to fracturing of embryo parts. The degree of mechanical injury depends upon cultivar susceptibility, seed moisture content, and machinery used. Seeds that are too dry become brittle. Wet seeds suffer from impact damage on embryos. Baldhead or snakehead in lima bean seedlings is caused by mechanical impact damage on the embryo during threshing. The seedlings may lack a growing point, radicle, or cotyledon(s).[8] Machines with high cylinder speeds result in damaged seed coats. Seed-coat injuries provide entry points for invasion of pathogenic and saprophytic microorganisms. Injured seeds are more susceptible to the phytotoxic effects of fungicide seed treatments, especially mercurials. Severely injured seeds may fail to germinate or may produce weak seedlings.

III. ENVIRONMENTAL EFFECTS

Humidity and temperature are environmental factors which have the greatest effect on nonparasitic disorders.

A. Temperature

In general, moist seeds are more sensitive to extremes of low rather than high tem-

FIGURE 33.   Severe mechanical damage on soybean (soyabean) (*Glycine max*)
seeds. (From Sinclair, J. B., Ed., *Compendium of Soybean Diseases,* American
Phytopathological Society, St. Paul, Minn., 1982, 49. With permission.)

peratures, with the degree of sensitivity depending upon seed moisture content. Dry
seeds usually can tolerate low temperatures (−185 to −192°C).[9] Frost damage to seeds
in the field has been reported in temperate countries. Cold-damaged rape seeds become
brown and shrunken.[10] Frost can injure developing seeds of cauliflower,[11] which may
fail to germinate or, if they germinate, produce weak, damaged seedlings.[9] Heating of
seeds during storage because of metabolic activities may result in seed damage, resulting in the production of abnormal seedlings.[9]

B. Humidity

Harvesting of moist seeds or staking a harvested crop in the field can maintain or
raise seed moisture content, which results in softening or cracking of seed coats. These
conditions favor development of storage fungi. Under humid conditions, *Raphanus
sativus* seeds develop a gray discoloration due to swelling of the subepidermal seed coat
parenchyma which results in distortion or cracking of the epidermis. This facilitates
invasion by *Alternaria* spp.[12,13] Desiccation of soybean seeds during maturation results
in cracking and wrinkling of the seed coat. Such seeds have a lower emergence than
nondamaged seeds.[14] Hollow heart, which is one of the most important physiological
disorders of pea seeds, is due to sudden drying of immature seeds at high temperatures.[15,16] Hollow heart symptoms appear as cavities or depressions on the adaxial face
of cotyledons. It is a common disorder of wrinkled pea seeds and peas having compound starch grains (Figure 34).[17,18] The cavities contain cells with reduced contents.
The hollow heart symptoms are determined by soaking pea seeds in distilled water for
16 to 18 hr at 20 to 25°C, followed by splitting the cotyledons apart and storing them
for 24 hr at 20 to 25°C. The cavities become conspicuous because they lose water
during drying and take more time to recover during imbibition.[17]

IV. MINERAL DEFICIENCIES

Mineral deficiencies in soil may result in the poor plant growth and thus poor quality
seed in the form of reduced seed size, abnormal seed shape, and low viability.

Nitrogen deficiency — Soils having low nitrogen and high potash and phosphorus
result in the nitrogen-deficiency symptoms in cereal grains. Nitrogen deficiency in
wheat is known as yellow berry. Symptoms appear on grains as light-yellowish spots
covering all or part of the seed. Such seeds germinate normally but may result in poor
market value. Affected grains have a high starch content, but are deficient in protein.
Nitrogen deficiency is corrected by nitrogen application.[9]

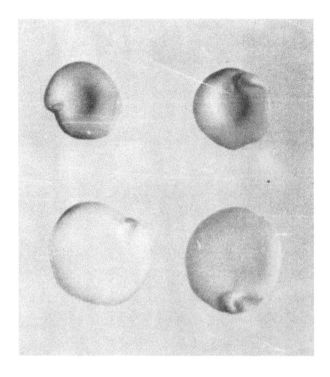

FIGURE 34.   Split pea (*Pisum sativum*) seed of cultivar CG 141
(above) showing cotyledons affected by hollow heart disease compared to nonaffected seed (below).

Manganese deficiency — Manganese-deficiency symptoms have been reported in
broadbean, common bean, pea, and many other crops. However, symptoms of manganese deficiency are more apparent in peas. The disorder, known as marsh spot, is
common in England (Figure 35).[19] Symptoms appear as discolored lesions in the center
of the adaxial face of the cotyledon.[17,20] Marsh spot is common in cultivars which
possess simple and large starch grains in the cotyledons. Affected cells secrete a pigmented material into enlarged air spaces due to the depletion of protoplasmic contents.[17] Seed coats of severely affected peas are brownish with sunken spots. Affected
seeds germinate, resulting in weak and poorly developed seedlings. Two sprays of 1%
manganese sulfate — first, just after the close of flowering and second, about 3 weeks
later — reduce incidence of this disorder.[21] An application of manganese sulfate to soil
at 100 to 200 kg/ha at flowering also controls marsh spot.[22,23] The incidence of marsh
spot in a seed lot is determined by soaking pea seeds in distilled water for 16 to 18 hr
at 20 to 25°C. The cotyledons are split apart and examined immediately.[17]

Potash deficiency — Potash deficiency in cucumber results in tapered seeds.[24] Marsh
spot peas, primarily due to manganese deficiency, were also reported due to the potash
deficiency.[25]

Boron deficiency — Boron deficiency, which causes hollow heart of peanut in Florida, results in a discoloration and rotting of seeds.[26]

V. INSECT DAMAGE

Insects cause distinctive damage to seeds and can be easily detected either by the
presence of the insects or damage to the seeds. Insects such as lygus bugs (*Lygus campestris, L. elisus, L. hesperus, L. oblineatus,* and *L. sallei*) cause injuries to carrot,

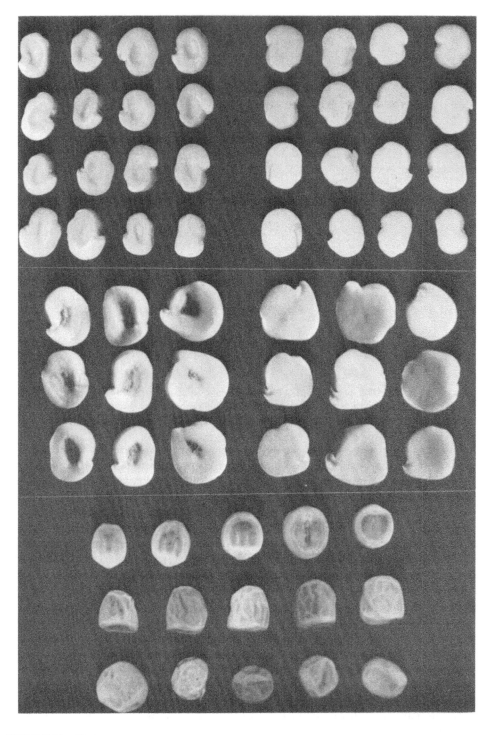

FIGURE 35.   Hollow heart and marsh spot symptoms on pea (*Pisum sativum*) seeds. Top: split seeds of cultivar Dark Skin Perfection showing hollow heart-affected cotyledons (left) and nonaffected cotyledons (right); middle: split seeds of cultivar Koroza showing marsh spot-affected cotyledons (left) and nonaffected cotyledons (right); and bottom: seeds of cultivar Mingomark showing squarish (top), conical (middle), and irregular (bottom) shapes. (From Singh, D., *Seed Sci. Technol.*, 2, 443, 1974. With permission.)

celery, coriander, and other crop seeds. The embryo is replaced by a cavity and the seed coat and endosperm remain undamaged. Such seeds appear normal but fail to germinate.[3] *L. hesperus* and *L. elisus* cause pits on lima bean seeds where the initial injury appears as watersoaked areas around a small hole, which shrivels to give an irregular pit that enlarges as the seeds mature. Such seeds may not attain normal size and appear shriveled. Pitted areas vary from a tiny, sunken pinpoints to large, irregular, crater-like yellow or brown spots in cotyledons. The cavities are filled with a brown, granular mass of necrotic cells and starch grains, the latter of which are particularly evident before the seeds dry.[27]

Insect injury to seeds often is followed by invasion of microorganisms or is accompanied by the introduction of a pathogen. Careful observation is required to determine the damage caused by the insect and by an associated pathogen.

REFERENCES

1. Kommedahl, T., Christensen, J. J., Culbertson, J. O., and Moore, M. B., The prevalence and importance of damaged seed in flax, *Minn. Agric. Exp. Stn. Tech. Bull.*, 215, 1, 1955.
2. Baker, K. F., Seed transmission of *Rhizoctonia solani* in relation to control of seedling damping-off, *Phytopathology*, 37, 912, 1947.
3. Baker, K. F., Seed pathology, in *Seed Biology*, Vol. 2, Kozlowski, T. T., Ed., Academic Press, New York, 1972, 317.
4. Harlan, H. V. and Pope, M. N., Some cases of apparent single fertilization in barley, *Am. J. Bot.*, 12, 50, 1925.
5. Lyon, M. E., The occurrence and behavior of embryoless wheat seeds, *J. Agric. Res.*, 36, 631, 1928.
6. Crill, P., Burgis, D. S., Jones, J. P., and Strobel, J. W., The fruit pox and gold fleck syndromes of tomato, *Phytopathology*, 63, 1285, 1973.
7. Stevenson, W. R., Tigchelaar, E. C., and Jackson, A. O., Corky stunt: a genetic disease of tomato, *Phytopathology*, 66, 132, 1976.
8. Borthwick, H. A., Thresher injury in baby limabeans, *J. Agric. Res.*, 44, 503, 1932.
9. Neergaard, P., *Seed Pathology*, Vol. 1 and 2, Macmillan, London, 1977, 1187.
10. Pape, H. and Harle, A., Rapsschaden dürch Maifröste, *Mitt. Schweiz. Landwirtsch.*, 31, 3, 1943.
11. Stapel, C. and Bovien, P., Mark-froafgrodernes sygdomme og skadedyr, *Det Kgl. Danske Landhusholdningasselskab*, Copenhagen, 1943, 234.
12. Jorgensen, J., Nogel underosgelser over ärsagerne til gräfarvningen af frø af gul sennep *(Sinapsis alba)*, *Statsfrøkontrollen København*, Beretning for det 96. Arbejdsär fra 1 Juli 1966 til 30 Juni, 1967, 78, 1967.
13. Neergaard, P., *Danish Species of Alternaria and Stemphylium*, Einar Munksgaard, Copenhagen, 1945, 560.
14. Moore, R. P., Wet, then dry, means poor soybean germination, *N.C. Agric. Exp. Res. Farming*, 18, 12, 1960.
15. Allen, J. D., Hollow heart of pea seed, *N. Z. J. Agric. Res.*, 4, 286, 1961.
16. Perry, D. A. and Harrison, J. G., Causes and development of hollow heart in pea seed, *Ann. Appl. Biol.*, 73, 95, 1973.
17. Singh, D., Occurrence and histology of hollow heart and marsh spot in peas, *Seed Sci. Technol.*, 2, 443, 1974.
18. Agarwal, V. K. and Gupta, R. K., Further reports of hollow heart in pea seeds, *Seed Res.*, 8, 75, 1980.
19. Furneaux, B. S. and Glasscock, H. H., Soils in relation to marsh spot of pea seed, *J. Agric. Sci.*, 26, 59, 1936.
20. Perry, D. A. and Howell, P. J., Symptoms and nature of hollow heart in pea seed, *Plant Pathol.*, 14, 111, 1965.
21. Koopman, C., The influence of manganese sulfate spraying on marsh spot of Schokker peas, *Tijdschr. Plantenziekten*, 43, 64, 1937.
22. Lewis, A. H., Manganese deficiency in crops. I. Spraying pea crops with solutions of manganese salts to eliminate marsh spot, *Emp. J. Exp. Agric.*, 7, 150, 1939.
23. Ovinge, A., Marsh spot in Schokker peas, *Tijdschr. Plantenziekten*, 43, 67, 1937.

24. Hoffman, I. C., Potash starvation in the greenhouse, *Better Crops Plant Food,* 18, 10, 1933.
25. De Bruijn, H. L. G., Kwade harten van de erwten, *Tijdschr. Plantenziekten,* 39, 281, 1933.
26. Harris, H. C. and Gilman, R. L., Effect of boron on peanuts, *Soil Sci.,* 84, 233, 1957.
27. Baker, K. F., Snyder, W. C., and Holland, A. H., Lygus bug injury of limabean in California, *Phytopathology,* 36, 493, 1946.

Chapter 12

DETECTION OF SEEDBORNE PATHOGENS

Seeds are complex, biologically dormant entities, containing a living embryo. In addition, microorganisms and/or viruses including parasites and saprophytes may be associated with them. Methods for detecting these seedborne microorganisms and viruses range from simple visual observation to sophisticated techniques such as electron microscopy and serology. Seed scientists have developed techniques for analyzing large numbers of seed samples for seed certification and plant quarantine. The PDC of the ISTA is active in standardizing some of these techniques for analysis of seed samples for certification.[1]

I. MAJOR OBJECTIVES OF SEED HEALTH TESTING

The major objectives of seed health testing are as follows.

Evaluation of the planting value of seed samples — A number of plant pathogens are transmitted through seeds (see Chapter 2). Detecting the presence or absence of microorganisms and viruses can help in predicting field performance of seed samples relative to emergence and subsequent disease development. Heavily infected or infested seed samples may be rejected for planting after seed health testing is completed.

Seed certification — Certification of seed samples relatively free of seedborne pathogens is gaining an importance as a criterion for high-quality seeds. At present, detection of only a few pathogens is part of a seed certification program in many countries. This is due to, in part, lack of proper documentation of the role of seedborne infections in disease development or unavailability of routine testing methods. Seed health testing method(s) have been standardized by ISTA through comparative seed health testing programs so that results are comparable between stations.

Seed treatment — Seed health testing can play a major role in determining whether or not a seed lot should be treated with a fungicide or other treatment based on the nature and degree of infection associated with the seeds. It helps to develop more precise seed treatments and also avoids unnecessary treatment of healthy seeds. With the introduction of systemic fungicides, seed treatment recommendations have become specific for crops and pathogens. In many countries fungicide seed treatment is done only on the basis of results from seed health testing. In India, wheat samples with up to 0.5% loose smut infection are certified without seed treatment and those with infections between 0.51 and 2% are treated by a seed-producing corporation with carboxin, a systemic fungicide.[2] Fungicide seed treatment based on seed health testing is required in Canada for peas and in Sweden for cereals. In Denmark, the Doyer filter-paper method, based on the ability of *Drechslera* and *Fusarium* spp. to attack and discolor rootlets of cereal seedlings, has been used for deciding whether or not fungicidal seed treatment should be applied to barley seeds.[3]

Quarantine — Seeds from germplasm collections derived from diverse agro-climatic zones are exchanged between countries to improve the local genetic material. However, with the exchange there are chances for introducing new pathogens or new races of indigenous pathogens, which can infect local crops. Thus, seeds are tested by international plant quarantine offices for dangerous or restricted pathogens, the entry of which are prohibited. The testing of seeds for quarantine can be complex because seeds often are treated in the exporting country before export. It is difficult to detect the infections in treated seeds by conventional methods. Moreover, seed treatment may not eliminate all pathogens associated with seeds. There are few standard techniques for removing fungicides from seeds to determine their fungistatic effect.

Advisability for feed or food — Fungal invasion by storage fungi can lead to toxin production associated with seeds of certain crops. Seed health testing may detect the presence or absence of toxin-producing fungi in seeds, which would help reduce the risk of using contaminated seed lots as feed or food.

II. CHOOSING SEED HEALTH TESTING METHODS

More than one method may be available for detection of a particular seedborne pathogen. The selection of a method depends upon the purpose of the test, i.e., whether the seeds are to be tested for seed certification, seed treatment, quarantine, etc. If for quarantine purposes, then highly sensitive methods are preferred because it is important to detect even traces of inoculum. In general, the method must be simple and quick and results must be reproducible. In addition, the results must be reliable with respect to field performance. The identifying characteristics of a pathogen should be recognizable with ease and certainty.

III. TESTING METHODS FOR SEEDBORNE FUNGI

The nature of seedborne fungi has been established for many fungi by demonstrating inoculum location within seed tissue using techniques which range from in vitro isolation from seeds to use of electron microscopy. The methods used for detection of seedborne fungi were reviewed.[1,4-7] The merits and demerits of the different methods are presented.

A. Examination of Dry Seeds

Certain fungi can be detected by direct observations of dry seeds or by using a stereobinocular bright-field microscope or hand lens to detect seed discoloration, morphological abnormalities, or fungal fruiting structures associated with the seeds (Figure 36). Seed infection can result in abnormalities in seed coat color, reduction in seed size and abnormal seed shape, necrosis, or production of fruiting structures. Such infections can be detected by direct observation. This method reveals partial to complete infections of certain seedborne fungi.

1. Seed Discoloration

Discoloration of the seed coat in certain crop seeds can be due to fungal infections (Figure 37). For example, the purple discoloration of soybean and guar seed coats indicates the infection of seed by *Cercospora kikuchii*. The whole seed surface or a portion of the seed coat may be purple.[8,9] *C. sojina* results in gray or brown discoloration of soybean seed coats.[10] Likewise, soybean seeds infected with *Macrophomina phaseolina* reveal indefinite black spots and blemishes on the seed coat (Figure 38).[11] Soybean seeds with a heavy infection of *Phomopsis sojae* appear discolored, fissured, flattened, and may be covered partially or wholly with a grayish-white mycelium,[8,12] whereas seeds infected with *Colletotrichum truncatum* have irregular grayish discolorations with minute black specks.[13,14] White-seeded *Phaseolus* bean seeds with olivaceous-green discoloration always are infected with *Alternaria alternata*. Seeds of common bean and pea with brown discoloration indicate infection by *Colletotrichum* sp.

Wheat seeds infected with black point or kernel smudge exhibit dark brown to black discoloration generally restricted to the embryo tip (Figure 39). The fungus most commonly associated with such seeds is *A. alternata* in addition to a rare occurrence of *Curvularia* spp., *Drechslera sorokiniana,* and *Phoma* spp. in India.[15,16] Black point symptoms of wheat are caused by several fungi including *Drechslera* sp. in North

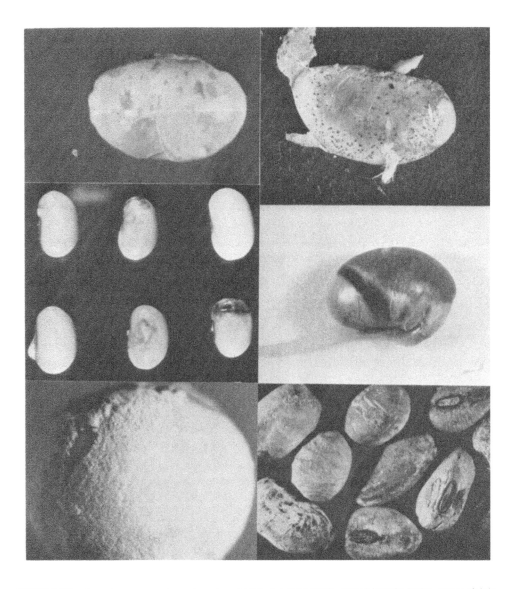

FIGURE 36. Symptoms on soybean (soyabean) (*Glycine max*) seeds caused by (clockwise, upper right): *Pseudomonas syringae* pv. *glycinea* (bacterial blight); *Colletotrichum truncatum* (anthracnose) — note presence of acervuli; *Nematospora coryli* (yeast spot); triple infection — *Cercospora sojina* (frogeye leafspot) (lower left quadrant), soybean mosaic virus (soybean mosaic) (dark streak), and *C. kikuchii* (purple seed stain) (upper two thirds); *Peronospora manshurica* (downy mildew) (encrusted with oospores); and *Diaporthe/Phomopsis* complex (pod and stem blight, Phomopsis seed decay). (All but *Colletotrichum truncatum* from Sinclair, J. B., Ed., *Compendium of Soybean Diseases*, American Phytopathological Society, St. Paul, Minn., 1982. With permission.)

America and *Alternaria* sp. in Europe.[17] Wheat seeds infected with *A. triticina* are shriveled and entire seed surfaces may be discolored.[18] Wheat seeds infected with *D. sorokiniana* are lightweight and smaller than normal and discoloration frequently extends to all seed parts. Also, infected seeds are shriveled and show fine ridges on the surface.[19] In black speck of rice, kernels may have dark brown to black spots which may spread over the entire surface. Such discoloration may be due to infection by *D. oryzae*, in addition to *Alternaria, Curvularia, Fusarium, Nigrospora,* and *Phoma* spp. and/or *Pyricularia oryzae*.[20,21] Red discoloration of rice kernels is caused by *Epicoccum nigrum*.[22]

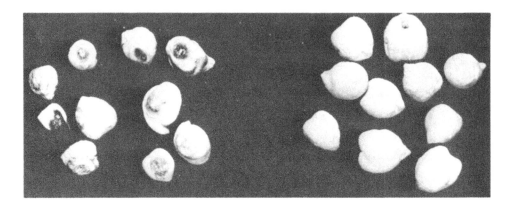

FIGURE 37.   Gram (*Cicer arietinum*) seeds showing symptoms of infection by *Ascochyta pisi* (Ascochyta blight) (left), and noninfected seeds (right). (Courtesy of M. V. Reddy, International Center for Agricultural Research in the Dry Areas, Syria.)

FIGURE 38.   Soybean (soyabean) (*Glycine max*) seed showing symptoms of infection by *Macrophomina phaseolina* (charcoal rot). Note microsclerotia in the tissues of the seed coat. (Courtesy of I. K. Kunwar.)

Septoria avenae, which causes black stem of oats, causes a green-black discoloration of portions of oat hulls.[23] *Cephalosporium acremonium* and *F. moniliforme* are associated with white streaks on maize kernels.[24,25] Wheat grains infected by the karnal bunt fungus are converted partially into a black powdery mass. Such symptoms indicate infection by *Neovossia indica.* Gray to brown wrinkled and badly deformed mungbean seeds carry heavy infections of *F. equiseti* and *Macrophomina phaseolina.*[26] Similarly, *Sclerotinia sclerotiorum (Whetzelinia sclerotiorum)* in beans results in chalky, discolored, and shriveled seeds.[27] Sunnhemp seeds infected by *F. equiseti, F. moniliforme, F. semitectum,* and *F. solani* are shriveled, discolored, and germinate poorly.[28] Mycosis spot disease of pea seeds, caused by *Ascochyta pisi, A. pinodella,* and *Mycos-*

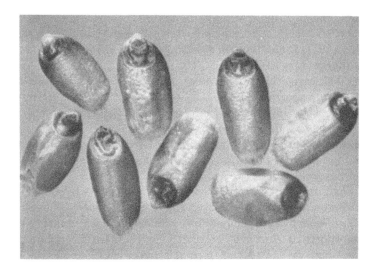

FIGURE 39. Nonspecific symptoms of black point of wheat (*Triticum aestivum*) caused by *Alternaria* spp. (From Hewett, P. D., *Seed Pathology — Progress and Problems,* Yorinori, J. T., Sinclair, J. B., Mehta, Y. R., and Mohan, S. K., Eds., Fundação Instituto Agrônomico do Paraná (IAPAR), Londrina, Brazil, 1979, 73. With permission.)

phaerella pinodes, shows necrotic yellow and brown spots.[29] Dark-colored sugar beet seeds are infested heavily with *Cercospora beticola*[30] Gray and black discolored seeds of *Eleusine coracana* probably are infected by *P. grisea*.[31] Soybean seeds affected by the yeast *Nematospora coryli* are wrinkled with light cream to white cheesy lesions.[32,33] A gray discoloration of *Datura* seeds is caused by *Alternaria crassa*.[34]

2. Morphological Abnormalities

Abnormal shape and reduced seed size can be caused by fungal infections. In stem gall of coriander, the fruits become partially or completely hypertrophied. Such fruits carry chlamydospores of *Protomyces macrosporus*. Alfalfa seeds infected with *Ascochyta imperfecta* are lightweight, dark, and shriveled.[35] *F. nivale*-infected wheat grains may be shriveled.[36] *Septoria nodorum* and *S. tritici* cause shriveling of wheat seeds.[37] Shriveled and medium-sized wheat seeds were infected with 10.2 to 16.6% *S. nodorum*, whereas plump, well-filled seeds were infected 7.7 to 8.1%.[38]

3. Mixed Fungal Fruiting Structures

Fungal fruiting structures, such as sclerotia of *Claviceps fusiformis* (ergot of pearlmillet) and of *Sclerotinia sclerotiorum* (white rot of legumes and sclerotial rot of cabbage) are found mixed with healthy seeds (Figure 12). Straw or other plant parts mixed with seeds during harvesting and threshing can carry plant pathogens. For example, *Melampsora lini* is carried with flax seeds in infected chaff.

The resting stages of several fungi are found as contaminants of soybean seed lots. Soil peds associated with cysts of the soybean cyst nematode (*Heterodera glycines*) (Figure 21) and sclerotia of *S. sclerotiorum* are not easily separated from seed lots during harvest or routine cleaning operations. The smut galls of *Melanopsichium pennsylvanicum* from *Polygonum pennsylvanicum* (smartweed) (erroneously referred to in the literature as soybean smut) have been found in seed lots in Australia and the U.S.[39,40] The smut galls and seeds of the weed host may be carried in soybean seed lots, and both the fungus and weed could become established in new areas in this way.

4. Observations Using a Bright-Field Microscope

Observing seeds under a stereobinocular microscope reveals the presence of chlamydospores, oospores, spores, etc. on seed surfaces. Confirmation of fungal species is done by spore morphology or culturing. Seedborne infections of the following diseases can be determined using this observations: wheat karnal bunt (*Neovossia indica*), wheat hill bunt (*Tilletia caries, T. foetida*), wheat dwarf bunt (*T. contraversa*), wheat flag smut (*Urocystis agropyri*), barley covered smut (*Ustilago hordei*), rice bunt (*N. horrida*), oat loose smut (*U. avenae*), oat covered smut (*U. hordei*), corn smut (*U. maydis*), corn and sorghum head smut (*Sphacelotheca reiliana*), sorghum loose smut (*S. cruenta*), sorghum grain smut (*S. sorghi*), sorghum long smut (*Tolyposporium ehrenbergii*), and pearlmillet smut (*T. penicillariae*). Pycnidia of *Phoma betae* (black leg of sugar beet), *Septoria apii* (leaf spot of celery), *A. pisi* (Ascochyta blight of pea), and *S. linicola* (pasmo on flax) have been observed on seed surfaces, as have been oospores of *Sclerospora graminicola* (pearlmillet downy mildew) on pearlmillet seeds. Safflower seeds carry teliospores of *Puccinia carthami*.[41] Soybean seeds carry oospores of *Peronospora manshurica* on the seed surface as a milky white to slightly brownish-red tinged grayish growth with the seed coat wrinkled and cracked (Figure 18).[42,43] Soybean cultivars differ in type, number of lesions, and production of oospore-encrusted seeds.[44]

5. Observing Seeds Under Near-Ultraviolet Light (NUV)

Observing seeds under NUV light without treatment can detect seed infection of some fungi. Fluorescence is an assumptive test and does not conclusively show the presence or absence of a pathogen. Pea seeds infected with *A. pisi* and *Stemphylium botryosum* exhibit yellow-green and dull orange fluorescence, respectively, under NUV (see discussion of light effects later in this chapter).

Examination of dry seeds often gives an indication of infection by some fungi, based on certain seed abnormalities. Seeds not exhibiting abnormalities may carry inoculum as internal hyphae without symptoms, and sexual or asexual spores. Therefore, negative results must be confirmed by additional tests.

B. Examination after Softening or Soaking Seeds

The method is useful for detecting fungal infections in which spores are liberated into water after soaking. Submerging cereal seeds in water or other fluids allows for liberation of spore masses from pycnidia. A slime layer on the water surface is examined for conidia. Infection of *Polyspora lini* in flax seeds can be determined using this method. The small hyaline, oval-shaped conidia of the pathogen can be detected at × 100 magnification. However, neither the presence of pathogenic hyphae nor spore viability can be determined. *Fusarium*-infected soybean seeds, when immersed in water, swell more rapidly than noninfected ones, since water penetrates easily through the injured testa. As the swollen seed increases to 1.5 to 2 times its original size, separation of infected seed is possible with nets or screens.[45]

C. Seed-Washing Test

This technique is used for detecting seedborne fungi carried as spores on seed surfaces. The method provides quick results, but is only useful for detection of those pathogens which adhere to the seed surface in the form of easily identifiable spores; therefore, it does not have a wide application. It reveals neither hyphal infection nor spore viability. If spores are within seeds or glumes and not on seed surfaces, they may not be detected.

A fixed number of seeds are placed in a flask containing detergent and sufficient water for soaking, then the flask is shaken on a mechanical shaker for 5 to 10 min.

The process is repeated if necessary. The suspension is centrifuged at 2500 to 3000 RPM for 10 to 15 min. The pellet is resuspended in water and examined under a bright-field microscope. The method is quick and can be adapted for detecting chlamydospores, oospores, smut spores, and conidia of *Alternaria, Cephalosporium, Curvularia, Drechslera, Fusarium, Peronospora,* and *Pyricularia*. The number of conidia are counted and the spore load per seed recorded. Kietreiber[46] advocated use of membranes or selection filters made from cellulose nitrate or acetate for counting *Tilletia caries, T. contraversa,* and *T. foetida* conidia in wheat seeds. The conidia are washed from seeds with hot alcohol or water containing a detergent. All conidia are at the same level and appear nearly the same as in water. A hematocytometer can be used to count conidia.[47] The washing test reveals the presence or absence of spores on seed surfaces, but not spore viability or hyphal infection. Neergaard et al.[48] compared the washing test and blotter method for estimating *P. oryzae* from rice kernels and found that conidia of the fungus often are dead or germinate poorly and did not produce growth on seeds using the blotter method.

The viability of conidia of *Curvularia, Drechslera,* and *Fusarium* can be tested by placing conidia in a drop of water on a glass slide in a humid chamber. Normally, conidia germinate within a few hours at 20 to 30°C. Hewett[49] proposed an infectivity test to test viability of *Septoria apiicola* conidia. Oospore viability of *Peronospora manshurica* can be determined by the tetrazolium test. Oospores are placed in a test tube containing 1 mℓ of distilled water in the dark for 48 hr at 30°C, then 1 mℓ of a 1% aqueous solution of 2,3,5-triphenyl tetrazolium chloride in buffer (pH 6.5 to 7) is added. The test tube is kept in darkness for 48 hr at 30°C. Oospores with cytoplasm that stains orange-red are considered viable.[43,50]

D. Incubation Methods

Two methods, blotter and agar plate, are recommended by ISTA for routine examination of crop seeds for fungal infections.[1] These are suitable for infections accompanied by hyphae, fruiting structures, or spores.[6] The tests are effective for detecting most seedborne fungi.[6,51-53] Identification is based on fungal morphology developed during incubation on the seed surface on blotters or on colony characteristics on an agar medium.

1. Blotter Method

The blotter test is a simple and inexpensive means of detecting pathogens and other microorganisms associated with seeds (Figure 40). The basic principle in this method is to provide a high level of relative humidity, and optimum light and temperature conducive for fungal development. A container capable of transmitting light, such as a plastic culture plate or zinc tray, is used in many countries. Blotters are soaked in distilled or tap water and placed in culture plates after draining off the excess water. Excess water stimulates bacterial growth that generally is antagonistic to fungal pathogens. Moistening the blotter with a solution of a wide-spectrum antibacterial agent or water at pH 5 reduces the bacterial growth. A fixed number of seeds are placed equidistant from one another. The number of seeds used depends upon seed size and type of infections. Large seeds, such as chickpea, maize, pigeonpea, sorghum, soybean, and wheat, are plated at 10 seeds per 9-cm culture plate while 20 seeds are used for smaller seeds, such as Brassica and pearlmillet. After placement of the seeds, culture plates are incubated at under NUV or fluorescent light with a 12-hr cycle light and dark at 20 ± 2°C. Seeds are usually examined within 8 days using a stereobinocular microscope. Fungal identification is based on mycelial growth, length, color, and arrangement of conidiophores; color, size, septation, and arrangement of conidia on conidiophores; and production and apperance of other fruiting structures such as acervuli, pycnidia,

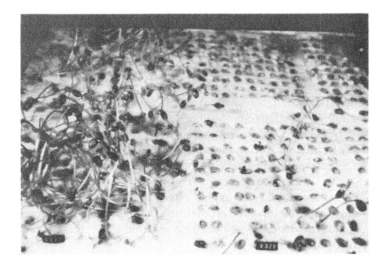

FIGURE 40.   Use of the blotter method to detect seedborne microorganisms associated with soybean (soyabean) (*Glycine max*) seed lots. Seed lot on the right is heavily infected with microorganisms. Seed lot on the left shows various seedlings with few or no seedborne microorganisms. (From Sinclair, J. B., Ed., *Compendium of Soybean Diseases,* American Phytopathological Society, St. Paul, Minn., 1982, 74. With permission.)

and sporodochia. Diagnostic characteristics of different species of *Colletotrichum, Curvularia, Drechslera,* and *Fusarium* on seeds have been reported in detail.[51-53] The blotter test provides conditions essential for the evaluation of the severity of infection on each seed and seedling, which may be important in epiphytology studies in the field.

Using the blotter method for certain crop seeds results in fast germination and lifting of the seed coats from the blotter surface. Most fungal structures are found associated with the seed coat and this seed coat displacement results in difficulties in evaluation and identification of microorganisms. To overcome this problem, 2,4-D (2,4-dichlorophenoxy acetic acid) in agar media is used. This was used first by Hagborg et al.[54] in testing bean seeds for the presence of *Colletotrichum lindemuthianum.* They recommend adding 50 μg/mℓ of 2,4-D to malt extract agar to inhibit seed germination without affecting fungal growth. Later, Neergaard[55] suggested dipping blotters in 0.2% 2,4-D for detecting fungal infections in cabbage seeds. Neergaard and Saad[56] reported an increase in detecting *Pyricularia oryzae* on rice seeds using this method. The commercial water-soluble salt formulation of 2,4-D is not toxic to fungi at the recommended rate.[56] Acid formulations of 2,4-D are toxic to *Phoma lingam* on crucifer seeds but the sodium-salt formulation is not.[57]

An alternative to 2,4-D blotter method is the deep-freeze blotter method advocated by Limonard.[58] Culture plates containing seeds are incubated for 1 day at 20°C, then at −20°C the second day, and then for 5 days under 12-hr dark and light at 20°C. The method is suitable for routine seed health testing of maize kernels.[59] Using this method, higher counts of *Cephalosporium acremonium* and *F. moniliforme* were recorded compared to blotter, rolled-towel, or agar-plate methods. In deep freezing, nutrients are freed from host cells and microorganisms are not injured. It is not suited for seeds such as bean and pea seeds, which become a putrefying mass. The test is used at the Seed Testing Station, Wageningen, The Netherlands, in routine testing for the graminicolous fungi, i.e., *D. avenae, D. dictyoides, D. siccans,* and *D. teres; Plenodomus lingam* in cabbage seeds; and *Alternaria* and *Stemphylium* in carrot seeds. It is easy, time saving, reliable, and gives higher counts compared to the blotter method.[60] The deep-freeze blotter method is superior to the 2,4-D blotter method also because the latter induces irregularly swollen or misshaped spores that are lighter in color. At higher than recommended concentrations of 2,4-D, the number of spores produced decreases.[58]

Fast-growing fungi may contaminate healthy seeds in close proximity. *Rhizoctonia solani* in jute seeds develops so rapidly that it causes secondary infection of adjoining seeds. Hence, reliable quantitative results are obtained by plating only one seed per culture plate.[61]

2. Agar-Plate Method

This method is used for identification of microorganisms associated with seeds based on growth and colony characteristics on a nutrient medium (Figure 41). Surface disin-festation is essential if a nonselective medium is used. Two nutrient agar media, malt-extract agar (MEA) and potato-dextrose agar (PDA), are suggested for seed health testing by ISTA.[1] Prior to plating, seeds are treated with a 1 to 2% sodium hypochlor-ite (NaOCl) solution to prevent saprophyte development. Usually 10 seeds per 9-cm culture plate are recommended. Incubation conditions are similar to those of the blot-ter method. Seeds are examined after 5 to 8 days incubation. Identification is based on macroscopic observations of the culture characters. MEA is used to detect seedborne fungi of cereals and flax in the Ulster method.[62] The presence of seedborne fungi on freshly harvested oat seeds results in an underestimation of *Pyrenophora avenae* using MEA. Treatment of seeds in a hot-air oven for 1 hr at 100°C before plating on 2% MEA and incubating for 6 days at 22°C is used to detect *P. avenae.*[63]

Using the agar-plate method, chances for development of superficial and/or fast-growing fungi are enhanced, and deep-seated, slow-growing types may fail to develop. Hence, selective media are used for isolation of some seedborne fungi and are more reliable than the blotter method. Miller et al.[64] found that oxgall-PDA is better than PDA alone for detecting *Fusarium* spp. because treatment of the seeds before plating is not necessary, and more seeds can be placed in the culture plate. Peptone-quintozene agar, used to detect soilborne Fusaria, also can be used to detect *Fusarium* spp. carried by seeds.[65] The choice of a particular medium for routine analysis of seeds depends upon the type of fungi associated with the seeds. In wheat seeds, if only *Fusarium* spp. are to be enumerated, then peptone-quintozene agar is suitable. However, when *Fusarium* and *Septoria* are to be enumerated, then oxgall-PDA should be used; and if *Drechslera, Fusarium,* and *Septoria* are to be enumerated, PDA with a treatment of seeds should be used.[66]

Selective media for certain seedborne fungi are as follows.

a. Fusarium culmorum

Malalasekera and Colhoun[67] developed a selective medium for isolating *F. culmo-rum* from wheat seeds which inhibits bacterial growth and most other seedborne fungi: 1 ℓ of Czapek-Dox agar is amended with 5 mℓ of an antibiotic, 6.25 mg captan, and 6.25 mg malachite green. Seeds are placed with the furrow downward in the center of each dish and incubated in the dark for 5 days at 25°C.

b. F. moniliforme

A selective medium for isolation of *F. moniliforme* from plant and soil tissues also was found specific for maize, pearlmillet, sorghum, and soybean seeds.[68] The medium is a modified Czapek-Dox agar containing 2 g $NaNO_3$, 1 g K_2HPO_4, 0.5 g $MgSO_4$, 0.5 g KCl, 0.01 g $FeSO_4$, 30 g sucrose, 0.5 g quintozene, 2 g yeast extract, 20 g agar, 0.05

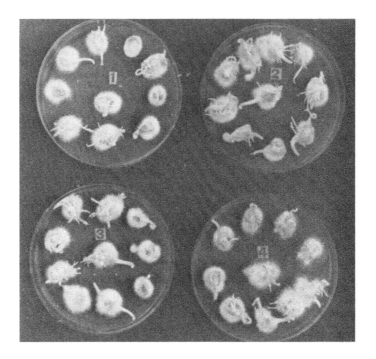

FIGURE 41.   Agar plate method used to detect *Fusarium moniliforme* in a selective medium on seeds of (1) maize (corn) (*Zea mays*), (2) sunflower (*Helianthus annuus*), (3) sorghum (milo) (*Sorghum vulgare*), and pearlmillet (*Pennisetum typhoides*) (above); and to detect a variety of seedborne microorganisms from surface-sterilized soybean (soyabean) (*Glycine max*) seeds on potato-dextrose agar (below). (Lower figure from Sinclair, J. B., Ed., *Compendium of Soybean Diseases*, American Phytopathological Society, St. Paul, Minn., 1982, 68. With permission.)

g fresh solutions of malachite green, and 0.75 g dicrysticin per liter. After autoclaving, the medium is poured into clean, sterilized culture plates. After 3 days, seeds are plated and incubated under NUV light with 16-hr light and 8-hr dark for 8 days at 25°C. Colonies of *F. moniliforme* appear within 5 to 8 days and are identified as spreading whitish-pink, fluffy mycelium on the agar surface with colonies surrounded by a pink-ish powdery growth with small, dot-like points. Reversed, the colony appears purple to bluish-purple.

c. Phoma betae

A highly selective method for the detection of *Pleospora björlingii (Phoma betae)* from sugar beet seeds was developed.[69,70] Seed clusters are nontreated or treated with 1% NaOCl, and five clusters of sugar beet seeds are spaced equidistant in each culture plate containing 1.6% water agar. After 7 days in the dark at 20°C, all seedlings and clusters are removed and inverted plates are examined at × 40 to 60 under a bright-field microscope for the presence of holdfasts. These structures appear as swollen branches at hyphal tips which later develop into a clump of swollen cells. While saprophytes normally develop on the agar surface, *P. betae* hyphae grow down into the medium to form holdfasts on the bottom of the agar. Later, a medium was developed which in-duces holdfasts or hyphal knots to turn brown to black, thus facilitating observation of the colonies:[71] 4 g K_2HPO_4, 1.5 g KH_2PO_4, 25 m*l* soil extract, 0.2 g boric acid, 0.1 g streptomycin SO_4, 0.1 g chlorotetracycline, 0.1 g benomyl, 10 g sucrose, and 17 g agar per liter of water. *P. betae* also can be detected on PDA after treatment with NaOCl.[72]

d. A. triticina

Nutrient agar is selective for *A. triticina* from wheat seeds:[73] 1.36 g K_2HPO_4, 1.06 g Na_2CO_3, 5 g $MgSO_4$, 5 g dextrose, 1 g asparagine, and 20 g agar per liter of distilled water. Seeds are surface sterilized in 0.1% mercuric chloride for 5 min followed by a sterile water rinse and are placed in culture plates containing the medium. Colonies of *A. triticina* appear white to olive-buff with a velvety cover of two to four conidia in chains after 10 days in continuous darkness at 25°C.

e. D. oryzae and Pyricularia oryzae

Fungi, in general, are known to produce a large number of metabolites differing in their nutritional, pH, and incubation temperature requirements, and resistance to toxic compounds.[74] Kulik[75] proposed the following method for detection of *P. oryzae* on a guaiacol agar (GA) medium: select 400 seeds at random, disinfect them in 1% NaOCl for 10 min, then place them on a GA medium — 5 g agar, 0.125 g GA, and 1 *l* water containing 0.5 g streptomycin sulfate or 0.8 m*l* lactic acid. Gently push seeds into the medium and incubate for 4 days in the dark at 22 to 25°C. Seeds having pale-pink to dark-red halos are counted as infected with *P. oryzae*. This medium also is effective for isolating *D. oryzae* from rice seeds. Prior to plating rice seeds on GA medium, immerse them in 1% NaOCl for 1 min, then in sterilized water for 1 min, and then place them on moist blotters for 24 hr at 22 to 28°C followed by 24 hr at 15 to −20°C. Colonies of *D. oryzae* are submerged, brick-red, and have a dendritic appearance. Colonies of *P. oryzae* produce a clean, dark-red halo in the medium adjacent to an infected seed. *A. padwickii* colonies may exhibit some brick-red color but are neither submerged nor dendritic. The GA method was found comparable to the blotter method (r = 0.95) and is less time consuming than checking individual seeds for *D. oryzae.*[75,76]

The method was found more sensitive than the blotter method for detecting of *D. oryzae.*[77]

3. Blotter vs. Agar-Plate Method

The blotter method is used widely for routine examination of seed lots for fungi which readily form mycelial growth on seeds. The method provides optimum conditions for a number of fungi, and one or more genera or species may develop simultaneously on a seed. The method requires examination of individual seeds and can be time consuming if seeds are heavily contaminated with saprophytes.

The agar-plate method is good to use to detect superficial infections of fast-growing fungi. It is used for cereal seeds in Sweden, flax and oat seeds in Northern Ireland, and pea seeds in Canada. A disadvantage of this method is that slow-growing or deep-seated fungi either fail to develop or their growth is suppressed by fast-growing types. de Tempe,[78] using agar plates, showed that fast-growing saprophytes adhering to seeds can be troublesome if one desires to detect internally borne, slow-growing pathogens.

The blotter method was found to give higher counts over the agar plate method in one study for the following fungi: *A. longissima, Curvularia lunata, D. oryzae, D. sorokiniana, A. padwickii,* and *Verticillium cinnabarinum* in rice seed; and *Colletotrichum truncatum, Curvularia lunata, F. semitectum,* and *Macrophomina phaseolina* in greengram. *C. pallescens* was the only fungus which appeared in a high incidence on agar.[20]

4. Factors Affecting Results of Various Incubation Methods

Identification of fungi using various incubation methods is based on color and rate of mycelial growth, sporulation, and spore morphology, all of which are influenced by growth medium, humidity, light and temperature, interfungal and fungal-bacterial interactions, length of incubation, and other characteristics. Factors influencing culture tests were reviewed by Leach[79] and Neergaard.[6]

a. Sampling

Often a small representative sample of a seed lot is tested and variation in samples can occur either in the field when a sample is collected or in the laboratory when a portion of a seed lot is taken for analysis. At either time, if samples are not drawn properly, variation in the results can be expected and may not give the true planting value of the seeds. It is recommended by ISTA that for routine testing at least 400 seeds should be tested.[1] The chance of variation is greater if seed samples contain large differences in seed size and morphology. It is desirable to repeatedly process the sample through a sample divider until the amount of seeds required is obtained.

b. Physical Condition of the Seed

Physical condition of seeds at testing is important since it influences the microflora. All factors which help in saprophyte development can pose a problem for growth of plant pathogens. Broken seed coats, high moisture content, etc. favor saprophyte development. A high frequency of saprophytes may overgrow deep-seated, slow-growing, pathogenic fungi. Proper seed handling is important to avoid unnecessary damage. In addition, seeds should not be harvested or threshed prematurely nor under high-humid conditions. This helps reduce seed invasion by storage fungi.

c. Storage of Seed

The microflora of a seed lot may change during storage. It is recommended by ISTA that storage conditions for seed samples required for later testing be cool and dry since humid and warm conditions favor development of storage fungi.[1] The storage period can influence the type and frequency of the seedborne fungi. Some fungi may be eliminated or reduced with storage. There are more chances for a decline in pathogens if seeds are stored at high temperature.

d. Incubation Containers

Several types of containers ranging from plastic culture plates to zinc trays are used for seed incubation. Leach[80] indicated that any container, including plastic or glass culture plates, capable of transmitting light can be used. At the Danish Government Institute of Seed Pathology, Copenhagen, plastic culture plates are used, and at The Netherlands Seed Testing station, Wageningen, zinc trays are preferred. However, secondary infections of healthy seeds from infected seeds may occur in either container. Limonard[58] suggested using 10 × 26 cm, 6-mm thick perspex plates with 75 holes of 15-mm diameter and a few millimeters deep. An individual seed is placed in each hole with a few drops of liquid so that seeds remain separated during incubation.

e. Incubation Media

In the blotter test, seeds are incubated on blotters soaked in sterile distilled water. Under standard, optimum conditions, blotters provide adequate relative humidity (over 90%) during incubation without addition of water. Blotter thickness and its capacity to retain water influences the relative humidity during incubation. Blotters should not dry fast nor be oversaturated to avoid the wet-blotter effect.[58]

PDA and MEA are used for seed health testing by ISTA.[1] The amount of media per culture plate influences the quality and intensity of transmitted NUV light.[81] This ultimately influences whether a colony is submerged or superficial. The number and type of colonies of *Pyrenophora avenae* produced on five media are comparable. However, those produced on a liquid malt-extract medium are submerged with less aerial mycelium and show a scarcity or delayed production of coremia.[82] Medium pH may influence the cultural characteristics such as the development of pigmentation.

f. Light

Light influences growth rate, colony morphology and color, sporulation, spore size, and morphology of many fungi encountered in seed health testing.[6,79,80] Fungi respond differently to different wavelengths. Seedborne fungi may respond to light in the following ways.[79]

(1) Stimulated Sporulation

Most seedborne fungi sporulate on exposure of an actively growing colony to UV wavelengths less than 340 mm. Light-sensitive fungi may sporulate normally when exposed to continuous light. For diurnal sporulators, a dark period stimulates sporulation. The NUV blacklight lamps emit longer wavelengths of UV radiation (300 to 380 nm) which are ideal for inducing sporulation in light-sensitive fungi.

Action spectra for light-induced sporulation were determined for conidia of *A. dauci* (less than 370 nm) and conidia and perithecia of *Pleospora herbarum* (less than 390 nm).[83] Leach and Trione's[83] single photoreceptor theory for sporulation may be involved in a number of fungi. Sporogen P_{310}, produced in a number of fungi in response to UV light, induces conidia and perithecia production in *P. herbarum*. The UV-light nutrition and genetic makeup of fungi determine P_{310} production. An isolate of *Ascochyta pisi* grown on Czapek-Dox agar required light for P_{310} production and sporulation. However, on Trione's medium, P_{310} is synthesized and sporulation occurs in the absence of light. Differences within species are expressions of different abilities to synthesize P_{310}. *Monilia fructicola* and *Stemphylium sarcinaeforme* sporulate well in darkness and produce P_{310} sporogens.[83] Diurnal sporulators (*Alternaria dauci, A. tomato,* and *S. botryosum*) probably have the same biochemical mechanisms of photosporogenesis, an inductive phase which leads to conidiophore formation, and a terminal phase which results in conidia formation. The inductive phase is stimulated and the terminal phase inhibited by NUV and blue light. The inductive phase proceeds efficiently at relatively high temperatures; the terminal phase at lower ones.[83] All diurnal sporulators produce substances similar to the P_{310} sporogens.[84]

Constant-temperature sporulators (*Drechslera catenaria, F. nivale,* and *Cercosporella herpotrichoides*) do not have distinct inductive and terminal phases, and have a lower optimum temperature range for sporulation. Sporulation in this group is abundant under continuous exposure to NUV light, and even more abundant when exposure is followed by darkness.[85] Different isolates of *Ascochyta pisi* may sporulate abundantly while others fail to sporulate in darkness on the same medium.[86]

Leach[80] recommend fluorescent blacklight lamps (NUV/360 nm) for seed health testing. Blacklight fluorescent tubes emit radiation in a continuous spectrum from 320 to 420 nm, which is suitable for seed incubation. The fixation of two 40-W tubes in a horizontal position 20 cm apart at 41 cm above the seeds is recommended. These tubes cover an area approximately 60 × 120 cm. An alternating cycle of 12-hr exposure to NUV light and dark throughout the test period may accommodate the diurnal sporulators as well as those that sporulate under continuous exposure.

A 12-hr light and dark cycle is recommended as a general standard procedure for testing seeds for fungal pathogens.[1] de Tempe[87] compared the incubation under NUV light and dark and found that an alternating cycle of 12-hr light and dark yields three times more *S. radicinum* and *Alternaria dauci* in carrot compared to continuous light. However, if NUV (360 nm) lamps are not available, cool, white daylight fluorescent lamps are used because they emit sufficient NUV to induce sporulation.[80,87] The incidence of *Colletotrichum truncatum* (blackgram, soybean), *Curvularia lunata* (rice, blackgram, greengram), *D. oryzae* (rice), *D. maydis* (maize), *Epicoccum purpuracens* (wheat, rice), *F. semitectum* (greengram, blackgram, rice), *Macrophomina phaseolina* (greengram, blackgram), and *Pyricularia oryzae* (rice) are practically equal at 20°C NUV, 28°C NUV, and 28°C daylight tubes.[20,88,89] Daylight fluorescent light is found as effective as NUV for detection of *Phoma lingam* in crucifer seeds.[57]

The seedborne fungi of rice, i.e., *C. lunata, D. oryzae, E. purpurascens, F. semitectum, Phoma* spp., *Pyricularia oryae, A. padwickii,* and *Ulocladium* sp. develop on rice seeds on blotters at 20°C under NUV or 28°C artificial daylight (ADL) but occur at a lower rate under complete darkness, particularly *Ulocladium* sp., while *A. padwickii* does not develop at all.[88] The use of NUV or ADL for 12-hr days at 20 to 28°C is suitable for most rice infections. The detection of *D. oryzae* decreases significantly under either continuous darkness or light. Most conidiophores are devoid of conidia under continuous NUV and ADL at 28°C.[89] The output of NUV radiation decreases with age in NUV lamps.[79]

Leach[79] suggested that daily exposure to NUV can be reduced to 6 hr and the recommended number of NUV lamps (2 × 40 W at 41 cm) can be reduced to a single lamp suspended 20 cm above the cultures. However, there is a need to work out the comparative effects of the two methods.

The NUV radiation may be harmful to eyes and skin. Radiation of 300 to 320 nm may produce skin cancer. Small levels of radiation are emitted by NUV black lamps. Direct exposure to NUV also may result in eye irritation which may last a day or more depending on the individual. Direct exposure of the eyes and skin should be avoided either by switching off the NUV lamps before entering the chamber or by wearing protective goggles and glasses.[91]

(2) Inhibited Sporulation

Sporulation may be inhibited by excessive levels of UV and the blue light UV at 380 nm. Pycnidia-producing fungi under an excessive UV light may produce pycnidia submerged in the agar. In diurnal sporulators, continuous light inhibits sporulation and the colonies are covered by conidiophores without conidia.

(3) Insensitive Fungi

Many fungi, particularly saprophytes, sporulate profusely in the presence or absence of light.

g. Temperature

Temperature influences in vitro growth, reproduction, and sporulation of fungi. The optimum temperature requirement for a majority of seedborne fungi ranges from 19°C (18 to 22°C) to 28°C (26 to 29°C).[6] Temperature also can influence colony color and conidia shape. Colonies of *E. purpurascens* appear light rose, wine-red, and yellowish-orange at 15, 22, and 28°C, respectively.[56] Kang et al.[90] found that for the major pathogenic fungi of rice, 20 and 28°C are equally effective under ADL tubes. *P. oryzae* gives higher counts at 28°C, *E. purpurascens* at 19°C, while *D. oryzae* grows equally well within a range of 18 to 30°C in rice seeds.[56] If a seed is infected by two pathogens of different temperature requirements, incubation at two temperatures may be used, as in the soil test, for the detection of *F. nivale, D. sorokiniana,* and *D. avenae.*[92] A majority of fungi grow and sporulate at a wide range of temperatures, and only a few are highly temperature sensitive.

h. Humidity

In blotter and agar tests, relative humidity and free water influence fungal growth and sporulation. Most fungi require a relative humidity above 90% for growth and sporulation, which is produced in most containers. However, available free water may influence results. Free water favors development of bacteria which antagonize fungal pathogens. The higher the free-moisture content of the container, the lower the percentage of detected fungi, especially in the case of small seeds that are enveloped easily by a water film and quickly absorb water. This is called the "wet blotter effect". Infection of *S. radicinum* in dry and wet blotters was 29.1 and 12.5%, respectively, on testing 50,000 carrot seeds; for *A. dauci* in 20,000 carrot seeds on dry and wet blotters it was 19.1 and 13.4%, respectively. Flax seeds, due to the presence of a gelantinous layer, absorb water quickly. The recovery of *Botrytis cinerea* was 15.7% in dry and 8.5% on wet blotters. Larger seeds are not surrounded by a water film and water uptake is slow. Thus, larger seeds are not affected by the wet blotter effect, except if seeds are soaked in water before plating.[60] Fungi developing on wet blotters show a poorer growth with less fungal structures than those developing on dry ones. Blotters are prepared by saturating them with water, then allowing them to drip for a few moments and then placing them into trays. Dry blotters are prepared by pressing the wet blotters between dry ones. The amount of water in dry, normal, and wet blotters is 30, 45, and 60 m*l*, respectively, or 135, 205, and 273% of blotter weight.[58]

To maintain a standard amount of water, the wicking arrangement, size, grade and make of filter paper, and amount of water to be added at the beginning of the test for different crop seeds, is standardized. Addition of water during incubation is avoided as far as possible because this would spread contaminants from one seed to another.

i. Pretreatment

Some seeds are heavily contaminated by a number of saprophytes, which hamper the development of slow-growing, deep-seated pathogens. Therefore, a mild disinfectant, such as a sodium hypochlorite solution (NaOCl) containing 1 to 2% available chlorine, is recommended for bean, cereal, and pea seeds when using the agar-plate method.[1] Treating soybean seeds in 0.5% NaOCl solution for 4 min is effective for detection of *Phomopsis sojae* from soybean seeds before incubation on PDA for 5 to 7 days at 25°C.[93]

Pretreatment of seeds reveals true infection counts of pathogens by eliminating the surface inoculum. A close relationship between laboratory estimation of *D. teres* in barley and field performance of barley seeds was found on testing pretreated seeds.[94] The pretreatment of pea seeds with NaOCl with 1% available chlorine from 30 sec to 40 min was found equally effective to using the agar test to detect *Ascochyta pisi* and *A. fabae*.[95]

Pretreatment does not selectively reduce only saprophytes; it can reduce the recovery of *D. maydis* and *F. moniliforme* in maize kernels.[59] Similarly, treatment with NaOCl reduces the incidence of *Colletotrichum truncatum*, *Curvularia lunata*, *F. semitectum*, *M. phaseolina*, and *Phoma* spp.[20] Several other disinfectants, such as ethyl alcohol, formalin, H_2O_2, and $HgCl_2$, can be used, but NaOCl is most satisfactory.[58] Sodium hypochlorite is harmless, not unpleasant to work with, and leaves no residue. Working solutions of NaOCl always should be fresh. Disinfection is carried out by immersing seeds in a 1 to 2% NaOCl solution for up to 10 min, rinsing them with sterile water, and then plating the seeds onto an agar surface using a flamed forceps.[58]

Yorinori and Sinclair[96] found that the highest recovery of *Cercospora kikuchii, C. sojina,* and *Phomopsis* spp. from threshed soybean seeds was obtained with no surface disinfestation; recovery was reduced with a 2-sec ethanol (86%) dip, followed by a distilled-water wash, then a 4-min soak in sodium hypochlorite (0.5%), another distilled-water wash, and placement on blotter paper or PDA. They concluded that the rate of transmission of seedborne fungi in soybeans can be underestimated when seeds are disinfested.

j. Length of Incubation Period

Incubation for 7 days is recommended for the blotter method.[1] However, an increase in the incubation period may result in higher counts of some fungi, such as *D. maydis* in maize.[59] Similarly, *Pestalotia guepini* on sorghum appears only after 15 days.[97] Derbyshire[98] emphasized the importance of incubating tomato seeds for 3 weeks to ensure full development of pycnidia of *Didymella lycopersici*. Ram Nath et al.[99] reported that *F. moniliforme* in rice seeds increased when the incubation period increased to 15 days. Mathur and Neergaard[88] found a higher percentage of *Alternaria padwickii* and *A. longissima* in blotter tests at 13 compared to 7 days at 20°C.

k. Amount and Type of Inoculum

The amount and location of inoculum associated with seeds varies among seeds and thus influences colony development in agar plates, and the quantity of conidiophores and conidia produced in the blotter test. In the blotter test, total fructifications of a fungus on a seed vary from a single spore or hyphal thread to complete coverage of the seed. In the agar-plate method, colony diameter around seeds can be influenced by the amount of inoculum. Whether a fungus is fast- or slow-growing also influences fungus growth on seeds. Location of infection in the seed influences growth. For example, if only the embryo is infected, fungus growth may develop more prominently on the germ end. Using the agar-plate method (Figure 41), colony diameter of *F. culmorum* in wheat depended upon spore load and ranged from 2.9, 5.4, 6.0, 6.5, and 7.2 cm with 10^2, 10^4, 10^5, 10^6, and 10^7 spores per seed, respectively.[67]

l. Interfungal and Fungal-Bacterial Interactions

Seeds can carry a range of microflora including pathogenic or nonpathogenic actinomycetes, bacteria, fungi, nematodes, and viruses. Interactions may occur within this complex.[60] These interactions may be among fungi or between fungi and bacteria. In testing for pathogenic fungi, other microorganisms usually influence this development. Antagonism among fungi is more common in agar tests, and between fungi and bacteria in blotter tests, especially with soybean seeds. However, bacteria play an important role both in the agar and blotter tests. In the agar-plate method, bacteria may grow on the seed surface and prevent fungal development. The addition of an antibiotic to agar media checks bacterial development. The problem is more difficult using the blotter method if seeds are heavily contaminated by bacteria and/or saprophytes. Both may be antagonistic to pathogen development, making identification difficult. The recovery of *F. moniliforme* from maize kernels originally with 60% infection was reduced to 9% with the application of dry spores of *Chaetomium globosum*, or to 19% using a suspension of *Bacillus subtilis*.[100]

E. Seedling-Symptom Test

This method involves raising seedlings under controlled conditions to allow for symptom development due to seedborne infections. It can predict field performance and estimate the number of infection foci per unit area in relation to seedborne seedling and plant diseases. It is used for postentry quarantine control, detection of certain obligate pathogens or pathogens that cause symptoms only on seedlings, and testing the efficacy of seed treatment fungicides. Seeds are sown in soil, sand, or other autoclavable growth media under optimum conditions. The seedling-symptom test is conducted using one of the following tests.

1. Rolled Paper Towel Test

This method generally is used for measuring seed germination in seed testing laboratories. Pathogens that cause seed decay, seedling blight, or seedling abnormalities can be detected. Marshall[101] found *Ascochyta* spp. associated with weak, decayed pea seedlings with stunted primary roots. Similarly, de Tempe[102] found that decay of flax seedlings was caused by *Alternaria linicola* or *Botrytis cinerea*. Hewitt[103] reported that carrot seeds infected with *A. dauci* produce seedlings with severe rotting. Wheat seedlings from *Septoria nodorum*-infected seeds develop malformed coleoptiles with brown spots and swellings (Figure 42).[104] Guerrero et al.[105] found *A. padwickii, Drechslera oryzae, Gibberella zeae,* and *Pyricularia oryzae* in decayed shoots and roots of rice.

Seeds are placed between layers of filter paper, blotters, or paper towels which then are placed on germinator trays in incubators or on room-type germinators with a relative humidity between 90 to 95%. The papers must be free from injurious chemicals and water-soluble dyes.[1] The towels are rolled and placed flat or in an upright position. Moistened porous paper or absorbent cotton or butter paper may be used as a base for the papers. Specificiations for incubation conditions for different crop seeds have been prescribed.[1] Normal seedlings, abnormal seedlings, ungerminated seeds, and hard seeds are recorded.

2. Blotter Test

Doyer[106] first used filter paper for identification of seedborne fungi based mainly on the symptoms. She used it to detect *Fusarium* spp. on cereal seedlings. *M. phaseolina* can be detected in sunflower seeds. It often results in the death of emerging radicles; discoloration of roots, hypocotyls, and cotyledons; or it can cause death of sunflower seedlings in the field.[107] *A. zinniae*, which causes seedling blight and lesions on sunflower cotyledons also can be detected.[108] Seedborne *Cochliobolus sativus* causes root discoloration during germination of barley seed on blotters.[109]

Phoma lingam in Brassica seeds is detected by symptom development and pycnidia

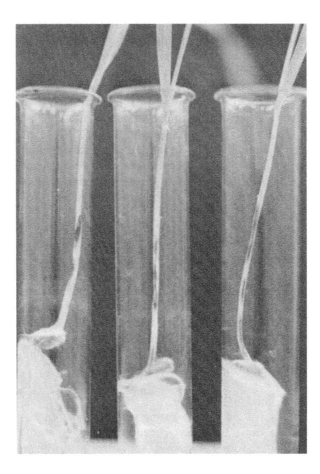

FIGURE 42.   Symptoms caused by *Septoria nodorum* (leafspot)
on coleoptiles of wheat (*Triticum aestivum*) seedlings from infected seeds grown on water agar test tubes. (From Khare, M. N.,
Mathur, S. B., and Neergaard, P., *Seed Sci. Technol.*, 5, 613,
1977. With permission.)

formation on seedlings. Two hundred seeds are placed on filter paper in refrigerator
box (10 × 20 cm, 6-cm high) and incubated in 12-hr light and dark for about 14 days at
6 to 7°C. Seeds with visible pycnidia then are plated in another box and incubated at
20°C. Infected seedlings become grayish-brown with a watery necrosis and pycnidia
form along the seedling axis.[110] This test is used to predict disease occurrence in the
field for barley.

Barley seeds sown on blotters are incubated for 7 days before being transplanted into
greenhouse soil. Seedlings develop symptoms of *Pyrenophora teres* and *P. graminea*
after 2 to 3 weeks.[111] A modification of this method was used in Denmark. Wheat
seeds were placed on moist filter paper in plastic culture plates for 3 days at 12°C, then
for 4 days at 20°C. *D. sorokiniana F. avenaceum*, and *F. nivale* caused discoloration
of rootlets.[3]

3. Agar Test in Test Tubes

This method was used to detect and demonstrate the seedborne nature of *Colletotrichum graminicola* in sorghum, *Didymella bryoniae* in cucumber and pumpkin, *Drechslera graminea* and *D. teres* in barley, *D. nodulosa* and *Pyricularia grisea* in *Eleusine
coracana*, *M. phaseolina* in sesame, and *S. nodorum* in wheat.[112]

Testing for *S. nodorum* in wheat can be done in test tubes (160 × 16 mm), or plastic microculture (2 m*l* capacity) or culture (9 cm) plates using 1% water agar. For test tubes, place one seed on a test tube slant with 10 m*l* agar, plug with a loose cotton plug, and incubate for 4 to 5 days. Remove the plug and incubate the tube for 14 days under alternating cycles of 12-hr light at 20°C. For microculture, fill microplate cavities with agar and place one seed in each cavity. Another microculture plate is reversed on top for 2 to 3 days, then removed. Place the microplate with seedlings in a polyethylene bag for 10 to 12 days. For culture plates, place 10 seeds on the water agar (10 to 20 m*l*) for 2 days, then remove the lids and place them in a polyethylene bag for 12 days. For all methods, seedlings are examined for *S. nodorum* symptoms, which appear on coleoptiles as small, light brown to black lesions, with a whitish-gray center. The seedling-symptom tests give counts similar to the deep-freezing blotter method.[112]

4. Soil Tests

Most incubation methods use artificial conditions for detection of seedborne fungi and may not precisely predict field performance of seeds. Hiltner[113] grew seeds under natural conditions for the estimation of seedborne infection in cereals. He used granulated (3 to 4mm) brickstone as a medium. Sterilized gravel, sand, soil, perlite, or vermiculite also can be used. One hundred cereal seeds are sown equidistant in a container and covered with 3 cm of brickstone chips and moistened with sterile water so that additional water is not required. After incubation in the dark for 2 weeks, seedlings are examined for root rot symptoms and fungus hyphae.

Soil tests using multipots are used at the Swedish Seed Testing Station to detect cereal seedling diseases caused by *Drechslera* spp., *F. nivale,* and *Septoria* spp., and seedling phytotoxicity from fungicide-treated seeds. Multipots are kept first for 2 weeks at 10°C to detect *F. nivale,* and then for 1 week at 20°C to detect *Drechslera* and *Septoria* spp. Infection by *D. graminea* cannot be determined since symptoms require more than 3 weeks to develop.[92] Soil tests in multipots closely predict field performance and efficacy of seed-treatment fungicides. Cottonseed infection by *C. gossypii* can be detected by planting the seeds in sterile sand in flats. The seeds are soaked in water prior to planting and incubated in a growth chamber under daylight with 85 to 90% relative humidity for 14 days at 25 to 28°C. Symptoms which develop on all seedling parts can be detected from seed lots with 1% infection or more.[114] *D. maydis* race T is detected in maize seeds by planting 100 seeds in sterilized soil in small plastic flats and placing them in the dark for 10 days at 18°C and then at 21,538 lx for 14 days at 23°C. Leaves of stunted seedlings show elongate necrotic lesions along the midrib, with seedling wilt and recovery of the fungus from the base of seedlings.[115] Symptom development depends upon inoculum within seeds; if there is a trace amount, no symptoms may develop. Symptoms develop only when infection is high, especially in pericarps.[115]

Seedling symptoms may not always be caused by a pathogen. When broadbean seeds are germinated, and if a high number of dead seeds are present, a large quantity of ammonia is given off by dead seeds. High levels of ammonia cause a black discoloration on seedlings, especially on root tips and leaves.[116] Major disadvantages of this test are that it normally does not distinguish between symptoms caused by different species within a fungal genus, it is difficult to recognize symptoms on older seedlings, seedling symptoms may not be characteristic of different genera, and seedling discoloration may be due to nonparasitic causes.

F. Fluorescence Method

The basic principle of this test is the ability of a fungus to produce a fluorescent substance under NUV light (a lamp of at least 1000 W). A *Septoria* colony associated

Table 12

COMPARISON OF THE APPEARANCE OF VARIOUS FUNGI IN SOYBEAN
SEED COAT TISSUES IN THIN SECTIONS PREPARED FOR BRIGHT-FIELD
MICROSCOPY[122,123]

Fungus	Hyphal width (μm)	Unstained mature hyphae	Hyphal aggregate	Mycelial mat	Sclerotia	Oil globule
Alternaria alternata	1.8—5.4	Brown	No	Yes	No	No
Cercospora kikuchii	1.3—3.0	Light brown	Yes	No	No	No
C. sojina	1.3—2.7	Brown	Yes	Yes	No	No
Colletotrichum truncatum	3—11	Brown	No	Yes	No	Yes[a]
Fusarium spp.	1.4—3.6	Hyaline	No	No	No	No
Macrophomina phaseolina	1.8—4.5	Brown	No	No	Yes	No
Phomopsis spp.	3.8—8.7	Hyaline	No	Yes	No	Yes

[a]   Prominent oil globules.

with wheat seeds or primary roots will fluoresce first pale blue, then yellow on filter paper. The yellow fluorescence can be seen on infected seedlings associated with hyphae or on a drop of exudate. Seven days are required for the test (the filter paper/freezing method): 3 days at 18°C, 3 hr at −20°C, and 4 days in the dark at 28°C.[117] *S. nodorum* colonies will fluoresce pale green to bright yellow-green on wheat seeds plated on oxgall agar containing 1.5% oxgall, 1% dextrose, 1% peptone, and 2% agar. Culture plates containing 10 seeds per plate are incubated under alternating 12-hr periods of light and dark. The plates are incubated with lids facing up for 3 days and then upside down for 3 days. Fluorescence is observed first in a few colonies after 2 days and in a maximum number after 5 to 6 days. Observations are made in the incubation chamber or room under the NUV lights. *S. nodorum* colonies appear round and orange-white under daylight. Dense hyphae cover seeds, coleoptiles, and roots. Reversed colonies are brown, surrounded by a light-brown halo. No bacteria or fungi associated with seeds show fluorescence similar to that of *S. nodorum*. Also, the percentage infection recorded is similar to that found when using the deep-freezing method. It is recommended for routine seed health testing.[118]

G. Histopathological Tests

Histopathology is used to locate inoculum in seed parts and obligate fungal infections. Maize seeds from plants systemically infected with *Sclerospora sorghi* show hyphae in the pericarp, endosperm, and embryonic tissues. Fungal hyphae are more concentrated in the coleoptile base and coleorhiza, but spores are more frequent in pericarp and endosperm.[119] *Phoma lingam* mycelium can be seen within seed coat tissues and occasionally in embryo tissues of *Brassica oleracea*.[120] Hyphae of *F. oxysporum* f. *carthami* are associated with xylem and sclerenchyma tissues in the pericarp and the parenchyma and sclerenchyma tissues of safflower seed coats. Although the fungus appears to develop from embryo parts on PDA, its presence could not be established in the embryo by histological methods.[121]

The following characteristics based on histopathological studies could serve as guidelines in distinguishing seven of the pathogenic fungi that colonize soybean seeds:[122,123] hyphal width, color of mature hyphae and the presence or absence of hyphal aggregates, mycelial mats, sclerotia, or oil globules in the hyphae. The fungi that can be distinguished are *A. alternata, Cercospora kikuchii, C. sojina, Colletotrichum truncatum, Fusarium* spp., *M. phaseolina,* and *Phomopsis* spp. (Table 12). If the fungus is hyaline, hyphae stain red or green with safranin, and light green; if lightly pig-

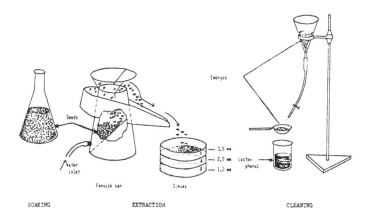

SOAKING   EXTRACTION   CLEANING

FIGURE 43. Diagrammatic representation of the embryo-count method for testing for infection by *Ustilago nuda* (loose smut) in barley (*Hordeum vulgare*) and other cereals. (From Yorinori, J. T., Sinclair, J. B., Mehta, Y. R., and Mohan, S. K., Eds., *Seed Pathology — Progress and Problems*, Fundação Instituto Agronômico do Paraná (IAPAR), Londrina, Brazil, 1979, 71. With permission.)

mented, the cell walls stain brown and the cytoplasm stains green or red. Hyaline hyphae stain blue with cotton blue; if lightly pigmented, the cell walls stain brown and the cytoplasm blue. If the hyphae are heavily pigmented, they will not take any of the stains.

H. Embryo-Count Method

The presence of loose smut fungi in barley and wheat seeds has been shown by various means. Fialkovaskoya[124] found that seed coats of severely affected seeds appear dull. Bubentozoff[125] detected the loose smut fungus in wheat seeds by isolating the fungus from seed pieces on potato-glucose agar medium (PGA). Seeds were disinfected with a $HgCl_2$ solution, soaked in sterile water for 24 hr, cut into pieces, plated on PGA, and incubated at $25 \pm 2°C$. This procedure yielded 60 to 75% infection. Wöstmann[126] observed loose smut hyphae in wheat seeds by microtomy. Naumova[127] reported that infected seeds exhibit a bluish fluorescence under UV light. However, none of these methods were adopted on a routine basis due to their unreliability. The only reliable method has been the embryo-count method which has been modified by a number of workers.[128-134]

Lactophenol is used for clearing and mounting fungi and plant tissues.[130,135,136] It is particularly useful for detecting loose smut hyphae in plant tissue and is especially useful in barley embryos and seedlings.[132] Skvortzoff[136] was the first to use aniline blue for staining hyphae of the loose smut fungus in wheat seed embryos. Morton[129] used boiling lactophenol to clear embryos and darken barley loose smut hyphae. Trypan blue (0.1 to 0.5%) was used by Boedijn[137] to stain fungi quickly. Popp[132] showed that 0.2% trypan blue stained *Ustilago tritici* hyphae in embryos. These techniques can be combined and modified for use in detecting loose smut infection in barley and wheat seeds.

The embryo-count method consists of four steps (Figure 43): extraction of embryos by soaking the seed in an alkali solution containing trypan blue stain, embryo separation, embryo clearing in lactophenol, and examination for the loose smut fungus hyphae. A simple embryo test was devised by Simmonds[134] and adapted for routine use at East Craigs by Laidlaw.[138] It is referred to as the Edinburgh method. Soak 120 g of barley seeds in 1 ℓ of a 5% solution of NaOH for 22 hr at 20 to 22°C. Transfer the sample to a nematode flotation vessel with a basal water inlet connected by tube to a hot water tap. The stream of hot water agitates the grain, allows embryos to float and pass over the lip with the water flow where they are collected into a series of graded sieves in tandem with 3.5-, 2.0-, and 1.0-mm perforations. Embryos are washed and dehydrated in 95% ethyl alcohol for 2 min, then transferred to a filter funnel and a lactophenol and water mixture is added. The embryos float, the chaff sinks, and embryos are collected. This process is repeated until clean embryo samples are obtained. Finally, embryos are placed in a 250-ml beaker with 75 ml anhydrous lactophenol and cleared by heating. Embryos submerged in lactophenol are examined in special grooved examination plates at × 18 to 25 using a moveable stage and substage illumination. Golden-brown hyphae of *U. nuda* are seen easily in the scutellum of infected embryos.[139] Pedersen[131] recommended that all embryos must be extracted since infected embryos are not easily extracted and thus can affect extraction rate. In contrast, Rennie and Seaton[139] and Hewitt[140] found no correlation between the embryo extraction rate and percent of loose smut infection. This method, with the addition of 0.02% trypan blue in soaking solution, can be used to detect infection by loose smut fungi in wheat seeds.[128,141,142] Agarwal et al.[128] simplified the technique by soaking seeds in a 5% NaOH and 0.02% trypan blue solution for 18 hr at 25°C. Embryos are recovered through flotation by adding 5% NaCl solution. The embryos then are washed in water, boiled in lactophenol, and examined with a binocular microscope.

Using the embryo-count method, a close correlation between seed infection and seed transmission was demonstrated in barley[129,131,133,138,143,144] and wheat.[130,132,134,141,142] In Scotland and Sweden, barley seeds are tested routinely for loose smut infection using the embryo-count method before seed treatment and distribution to farmers. At the Official Seed Testing Station, East Craigs, Scotland, detection of loose smut infection by the embryo-count method has been used since 1959.[139] Analysis of certified wheat seed lots for loose smut infection is part of the seed certification program of the Uttar Pradesh Seeds and Tarai Development Corporation, Ltd., India.[141]

The embryo-count method is used to detect *S. graminicola* hyphae in pearlmillet seeds. The hyphae were detected in the pericarp, aleurone layer, endosperm, membrane covering the scutellum, and scutellum. The fungus was not observed in the plumule bud or radicle. The hyphae appear as long, smooth, aseptate branced to net-like robust threads with forked haustoria.[50]

I. NaOH Seed-Soak Method

In rice bunt, caused by *N. horrida*, seeds are partially or completely replaced by a black powdery spore mass, which may not be visible. The NaOH seed-soak method can detect internal rice bunt infection in unruptured seeds. Rice seeds are soaked in 0.2% NaOH for 24 hr at 18 to 25°C. The solution is poured off and the swollen seeds spread over blotter paper to remove excess moisture. Discolored seeds are examined under a stereobinocular microscope. Jet-black seeds are infected and brown to dull seeds are not. Infection is confirmed by puncturing each discolored seed with a fine needle in a drop of water for release of bunt spores.[145] The same technique is effective in estimation of karnal bunt infection (*N. indica*) in chemically treated and untreated wheat seed lots.[146]

IV. TESTING METHODS FOR SEEDBORNE BACTERIA

Detection of seedborne bacteria is more difficult than that of seedborne fungi. Seedborne bacteria can be detected by isolation on agar media, seedling symptomatology, host infectivity, serologically, and by using phage plague. The lack of reliable, simple,

FIGURE 44. Bean (*Phaseolus vulgaris*) seeds photographed in UV light with four seeds on the right infected with *Pseudomonas syringae* pv. *phaseolicola* (bacterial blight) exhibiting fluorescense, and two nonfluorescent seeds on the left. (From Parker, M. C. and Dean, L. L., *Plant Dis. Rep.*, 52, 534, 1968.)

and quick method(s) is a limiting factor in the routine testing of crop seeds for a majority of the seedborne bacteria for seed certification. The methods used were reviewed by Guthrie.[147]

A. Visual Observations of Dry Seeds

Observing seeds visually or aided with light may indicate the presence of bacterial infection. Visual observation of dry seeds can give a quick indication of possible bacterial infection, but symptomless seeds may carry bacteria within them. Thus, the practicality of this test is limited. Rice seeds infected with *Xanthomonas campestris* pv. *oryzae* appear dark tan.[148] Glume blotch symptoms (*Pseudomonas syringae* pv. *oryzicola*) on rice florets in panicles appear as small, dark brown areas surrounded by green to light brown tissue to a dark brown discoloration of the florets.[149] Bean seeds showing 80% discolored hilums are infected with *X. campestris* pv. *phaseoli*.[150,151] Similarly, *X. campestris* pv. *phaseoli* causes darkened areas in infected dry bean seeds.[152] *Corynebacterium flaccumfaciens* pv. *violaceum* induces purple discoloration and shriveling in bean seeds.[153] In soybean *C. flaccumfaciens* pv. *flaccumfaciens* and a related nonpigmented *Corynebacterium* sp. were isolated from cream-colored soybean seed coats. The pigmentation usually spreads in streaks but at times may form an irregular halo around the hilum. Only the outer layer of seed coat cells are pigmented.[154] Pepper seeds infected with *C. michiganense* pv. *michiganense* show a brown discoloration and generally are smaller than noninfected seeds.[155]

Examination of seeds under UV light is useful in the initial detection of *P. syringae* pv. *phaseolicola* in bean seeds (Figure 44). A 100-W source emitting energy at a peak of 3650 Å distributed over a broad area is effective. The blue-white fluorescence is visible in bean cultivars with white seed coats or with white seed coats and colored eye spots but not with dark-colored seed coats. Fluorescence indicates that seeds are likely to be internally infected with the pathogen. It must be further confirmed by supplementary tests.[156] Parker and Dean[156] found that 25.5% of the seeds with fluorescence yielded *P. syringae* pv. *phaseolicola,* but 4.6% of the nonfluorescent seeds also possessed the bacterium. Taylor[157] and Wharton[158] also found that all fluorescing bean seeds may not carry the bacterium.

B. Isolation on Agar Media

Certain bacteria are capable of fluorescing, hydrolyzing starch, or growing on selective media, and these characters are used to isolate them from seeds, but the methods must be supplemented with pathogenicity, serological, or phage-typed tests. Differences observed among colony characters on agar media of different species and various isolates are not reliable.

X. campestris pv. *oryzae* is isolated from rice seeds on a nutrient agar media containing 0.5% glucose. Seeds are surface sterilized with 0.1% mercuric chloride. Yellow colonies of the bacterium develop from seeds after 24 to 48 hr at 30°C. Then single colonies are isolated from dilution plates and identified by their pathogenicity, morphological, cultural, and biological characters.[159] For detection of *X. campestris* pv. *zinniae,* zinnia seeds are plated on nutrient agar amended with 100 µg/ml captan and incubated for 3 days at 28°C. Individual colonies are transferred to nutrient agar and incubated for 3 to 4 days at 28°C. These cultures are used for confirmation of pathogenicity tests.[160]

For detection of *X. campestris* pv. *phaseoli* and *X. campestris* pv. *phaseoli* var. *fuscans,* 10,000 (1.6 kg) surface-disinfected bean seeds are ground wet and a dilution series prepared. A sample of each dilution is plated on agar plates. After 5 to 10 days, *Xanthomonas*-like colonies are isolated and phage typed and/or tested for infectivity on host plants. This method can detect either or both bacteria in 1 in 10,000 seeds.[161] For the detection of *P. syringae* pv. *phaseolicola,* dry bean seeds are ground to a flour and suspended in sterile tap water. Samples from a serial dilution of the supernatant are plated on a medium which allows *P. syringae* pv. *phaseolicola* to fluoresce. Confirmation is by serological test, bacteriophage, or host plant. This method can be used to detect other seedborne bacteria, such as *P. syringae* pv. *pisi* in pea and *X. campestris* pv. *phaseoli* in bean.[161,162] Schaad[163] used an indirect immunofluorescence test (IIF) for identifying *X. campestris* pv. *campestris* based on colony morphology on Sx media (10 g soluble potato starch, 1 g beef extract, 5 g ammonium chloride, 2 g potassium diphosphate, 1 ml of a 1% solution in 20% ethanol methyl violet B, 2 ml of 1% solution methyl green, 250 mg cycloheximide, 15 g agar, 1 l distilled water, pH 6.8).[164]

For IIF, bacterial cells are treated with an antiserum or Y globulin to 70 s ribosomes and stained with fluorescein isothiocyanate-conjugated antirabbit globulin. Stained cells are observed using fluorescence microscopy. Ribosomes are highly specific. The method is simple and rapid compared to pathogenicity tests.

Five selective media (D-series) have been developed for the recovery of species of *Agrobacterium, Corynebacterium, Erwinia, Pseudomonas,* and *Xanthomonas* from diseased tissues[165] and seeds. Hsich et al.[166] developed a medium for detecting streptomycin-resistant mutants of *X. campestris* pv. *oryzae* by touching seed abscission surfaces to a Wakimoto medium[167] supplemented with 250 µg/ml streptomycin sulfate. Saprophytic bacteria appear in 2 to 3 days, and the large and mucoid colonies of *X. campestris* pv. *oryzae* in 4 to 5 days. A selective medium was developed for isolation of *C. michiganense* pv. *nebraskense* from all maize plant parts.[168] The ingredients are added to 1 l of sterile nutrient broth yeast cooled to 50°C. The broth yeast consists of 25 mg malidixic acid (freshly solubilized in 0.1 *M* NaOH, 10 mg/ml), 32 mg polymyxin B sulfate, 40 mg cycloheximide, and 0.0625 ml Bravo 6 F.

X. campestris pv. *campestris* hydrolyses starch. This characteristic is used for isolating it from Brassica seeds. Schaad and White[169] used Sx agar.[164,169] Seeds are washed in a detergent solution for 20 min, rinsed in sterile water, and surface sterilized in 1% NaOCl. Seeds are incubated in sterile water for 20 hr at 30°C and after drying in air steam plated on Sx agar for 5 and 10 days at 30°C. Seeds surrounded by a gray to purple mucoid colony are a positive indication of *X. campestris* pv. *campestris*. Representative colonies are screened by streaking on yeast extract-CaCO₃ agar and counting yellow mucoid colonies. Identification is confirmed by a pathogenicity test. If seeds become overgrown by other bacteria, then a second 1000-seed sample is assayed by plating surface-disinfected seeds on Sx agar.

Starch hydrolysis by *X. campestris* pv. *campestris* for detection of the bacterium in Brassica seeds was used by Lundsgaard.[170] Brassica seeds are surface disinfected and groups of 100 seeds are soaked in sterile water in each of 10 culture plates for 24 hr at 22 to 24°C. Two drops of the water suspension are streaked on beef-peptone agar plus starch and incubated for 3 days at 28°C. Yellow colonies are tested for starch hydrolysis by adding several drops of Lugol's solution around the colonies. The medium around the colonies will remain unstained if starch was hydrolyzed, while the remainder of the medium will turn dark blue. Yellow colonies hydrolyzing starch and suspected to be *X. campestris* pv. *campestris* are identified using serological or infectivity tests.

A semiselective medium and a serological test (SSMS) was developed for detecting *X. campestris* pv. *phaseoli* in bean seed. The medium permits growth to high levels even in the presence of contaminating microorganisms. A serological test is used to confirm identification. Seeds are surface sterilized in 2.6% NaOCl for 2 min and rinsed with sterile distilled water, then placed on a semiselective medium (1 g yeast extract, 25 mg cycloheximide, 2 mg nitrofurantoin, 1 mg malidixic acid, 0.05 mg gentamicin in 1000 m*l* of 0.01 *M* phosphate buffer, pH 7.2) at 3 m*l*/g seed for 48 hr with shaking. Bacteria are sedimented by a 15-min centrifugation at 5000 *g* and resuspended in 1 m*l* sterile buffered saline (0.85% w/v NaCl in 0.01 *M* PO₄ buffer, pH 7.2). The suspension is steamed for 60 min at 100°C and tested by an agar gel double diffusion test. This method is more reliable and efficient than Saettler's technique.[171,172] It can detect an initial concentration of 400 cells per milliliter.[173]

C. Seedling Symptomatology Test

This test is used primarily to detect bacterial pathogens, but can be used for fungal pathogens. The seedling symptomatology test is used for the detection of *P. syringae* pv. *glycinea* (soybean), *P. syringae* pv. *lachrymans* (cucumber), *X. campestris* pv. *campestris* (cabbage), *X. campestris* pv. *malvacearum* (cotton), *X. campestris* pv. *oryzae* (rice), *X. campestris* pv. *vesicatoria* (pepper), and *X. campestris* pv. *zinniae* (zinnia). The incubation conditions of light and temperature and symptom development have been described for specific seedborne bacteria.[160,174-180]

P. syringae pv. *glycinea* is detected in soybean seeds by predisposing germinating seeds and seedlings to conditions favoring formation of water-soaked lesions on cotyledons. Seeds are soaked in sterile deionized water for 2 hr, then sowed in water-saturated vermiculite for 2 days at 24°C. Germinated and nongerminated seeds are shaken for 30 min with sterile, coarse sand. During shaking, seed coats become partly or completely detached, exposing the cotyledon surface to wounding. The seeds are replanted in vermiculite and incubated in 95% relative humidity and a 16-hr daylight at 25 to 27°C. Characteristic lesions develop on cotyledons which are dark green and water soaked. Older lesions consist of pitted necrotic areas with water-soaked edges on the outer lower surface of cotyledons.[174] The method is semiquantitative and infection levels may be overestimated due to contamination of healthy seedlings during shaking.

P. syringae pv. *lachrymans* is detected in cucumber seeds by symptoms on cotyledons of 8- to 10-day-old seedlings. This method is used in Israel.[175] Heavily infected seeds often fail to germinate, resulting in an underestimation of the level of seedborne inoculum.

X. campestris pv. *campestris* is detected in cabbage seeds by a growing-on test.[176] Five hundred seeds are germinated (2.5 seeds per square centimeter) on moist absorbent paper in a plastic box (22 × 14 × 8 cm) at 22°C. The bacterium causes progressive collapse, blackening, and death of seedlings in 8 to 18 days. Collapsed seedlings are macerated in water and pricked into leaf veins of Brassica indicator plants, which results in development of characteristic symptoms. Seedlings free of bacteria remain healthy for at least 21 days. The method does not allow an estimate of the percent of infected seeds. Srinivasan et al.[177] developed a technique for the routine detection of *X. campestris* pv. *campestris* in crucifer seeds. One thousand seeds are soaked for 3 to 4 hr in an antifungal antibiotic such as aureofungin (200 μg/mℓ) and then plated on 1.5% water agar at 20 seeds per 9-cm culture plate. The number of infected seedlings is recorded at 8 and 12 days after incubation in the dark at 20°C. Symptoms appear as V-shaped black rot lesions on cotyledons and true leaves. A positive infection is recorded by observing the bacterial ooze from vascular bundles through a cut of the lesion. The bacterium is isolated from diseased tissues and identified.

X. campestris pv. *malvacearum* is detected in cottonseeds by counting lesions of bacterial blight on cotyledons 21 days after planting. Seeds are planted in steam-sterilized sand in 10-cm clay pots under plastic-covered cages with 78 to 100% relative humidity.[178] Infection of *X. campestris* pv. *malvacearum* also can be detected on 14-day-old seedlings grown from cottonseeds under 85 to 90% relative humidity at 25 to 28°C.[113]

X. campestris pv. *oryzae* is detected in rice seeds using the paper towel method after 5 days at 30 ± 2°C. Paper towels are covered with a polyethylene sheet to maintain humidity. Small pieces from pale seedlings of poor growth with light to dark brown coleoptiles and discolored sheaths are examined under a microscope for bacteria streaming. Typical bacterial blight symptoms develop when bacterial ooze from the seeds is used as inoculum on susceptible seedlings using the clip method.[179]

X. campestris pv. *vesicatoria* is detected in chilli pepper seeds by planting them in pots with autoclaved soil. After 15 days, when seedlings are in the cotyledonary stage, pots are transferred to a humidity chamber with 95 to 100% relative humidity for 3 to 5 days at 24 to 30°C, then to a greenhouse at 20 to 30°C. Typical symptoms appear on cotyledons after 10 to 20 days as round, greenish, small spots which later become ash colored. Later, some lesions coalesce and cotyledons turn yellow and drop, proving pathogenicity.[180]

X. campestris pv. *zinniae* is detected in zinnia seed by densely planting seeds under warm and humid conditions in the greenhouse. Distinct lesions develop on cotyledons in 5 to 14 days.[160]

D. Infectivity Test

Indicator or susceptible host plants can be inoculated with seed washings and seed extracts for detection of some bacteria. The infectivity test is simple and reliable for detecting seedborne bacteria in groups of seeds, but does not provide the percent seed infection unless each seed is tested separately. The detection of *X. campestris* pv. *phaseoli* in bean seeds can be done by inoculating indicator plants with seed washings. Seeds surface sterilized in 2.6% NaOCl for 10 min are incubated in 10 g/ℓ of yeast extract broth for 24 hr. The liquid surrounding the seeds is used for inoculating the primary leaf node of susceptible seedlings, which develop typical symptoms.[171,172] Washings collected from cut soybean seeds and inoculated into healthy leaves can reveal the presence of *P. syringae* pv. *glycinea*. Soybean seed samples are assayed by macerating

10 g dry seeds in a Waring® blender for 1 min, then shaking them in 50 m*l* water for 1 to 3 hr. The washings then are forced into healthy leaves.[181]

For detection of *X. campestris* pv. *campestris,* 2000 cabbage seeds are incubated in 100 m*l* water for 12 hr at 27°C. The liquid is collected and used for inoculating 1-cm cuts in detached radish leaves. The leaves then are placed on a wet blotter in a covered plastic pan under 6456 lx at 27°C. Symptoms appear in 4 to 5 days as yellow margins around the wounds followed by blackening of adjacent veins. Isolates from diseased leaves are inoculated to other cruciferous plants.[182]

For detection of *C. michiganense* pv. *michiganense,* dry tomato seeds ground in Ringer's solution are filtered and used for inoculation of 3-week-old tomato seedlings. Plants are examined after 4 weeks for wilting and stem cankers. The method requires up to 7 weeks for completion.[183]

Venette[184] developed the Dome test for the detection of *P. syringae* pv. *phaseolicola* in bean seeds. The liquid collected from soaked bean seeds is vacuum infiltrated into bean seedlings. Disease symptoms appear as water-congested spots in leaf tissues after 7 to 10 days.

E. Serology

Attempts have been made to detect bacteria in seeds using serological methods. *P. syringae* pv. *phaseolicola* infection in bean seeds can be detected rapidly and accurately using serology. The microprecipitin test is more accurate and rapid than the tube agglutination or gel diffusion tests. Infection is detected by incubating surface-sterilized seeds in sterile distilled water for 36 hr after surface sterilization and then testing the supernatant with the antiserum. Four infected bean seeds per 45 kg (160,000 seeds) can be detected with 95% accuracy. Serological testing of bean seed to detect *P. syringae* pv. *phaseolicola* is routine in Idaho.[185,186] *X. campestris* pv. *manihotis* is detected in cassava seed embryos by the enzyme-linked immunosorbant serological assay (ELISA).[187]

F. Phage-Plague Method

The phage-plague method is used for detecting *C. michiganense* pv. *michiganense* in tomato,[188] *P. syringae* pv. *atrofaciens* in wheat,[189] *P., syringae* pv. *phaseolicola* in beans,[186,190] *X. campestris* pv. *oryzae* in rice,[191] *X. campestris* pv. *phaseoli* in bean,[161,192,193] and *X. campestris* pv. *vesicatoria* in tomato[188] with the use of specific bacteriophages. The basic principle lies in the increase in the number of specific bacteriophage particles in the presence of susceptible bacteria.

Katznelson and Sutton[193] first developed the phage method for detection of *P. syringae* pv. *phaseolicola* and *X. campestris* pv. *phaseoli* in bean seeds. Surface-sterilized seeds are ground in sterile nutrient broth and incubated for 24 hr to permit multiplication of the phage-sensitive bacterial cells. About 4000 to 5000 phage particles are added to 10 m*l* of the macerate in a flask. Samples (0.1 m*l*) of the mixture are spread on culture plates seeded with a suitable indicator bacteria. The number of zones of lysis (plagues) are counted. Another sample is plated 6 to 12 hr later. The increase in the number of lysis zones indicates the presence of the homologous *P. syringae* pv. *phaseolicola* or *X. campestris* pv. *phaseoli* in the seed sample. Sutton[189] isolated 68 phages for *P. syringae* pv. *atrofaciens* from cereal seeds. His phage "Pg 8" was used for detecting *P. syringae* pv. *atrofaciens* in wheat seeds using the method of Katznelson and Sutton.[193] A specific strain and a polyvirulent phage for *P. syringae* pv. *pisi* was isolated from peas. The polyvirulent phage is used in detecting *P. syringae* pv. *pisi* in pea seeds by the Katznelson and Sutton method.[193] Phages for *P. syringae* pv. *coronafaciens* and pv. *atrofaciens* were obtained from oat and wheat seeds, respectively.[190]

Homogenates of tomato seedlings from seeds are used to detect *C. michiganense* pv. *michiganense* and *X. campestris* pv. *vesicatoria* using the phage-plague count method. It was found that 3% of both bacteria occurred in seeds from naturally infected tomatoes, extracted by the fruit pulp formation method; while 62% and 43% of *C. michiganense* pv. *michiganense* and *X. campestris* pv. *vesicatoria*, respectively, occurred in seeds extracted with fermentation.[188] Multiplication of added phage may be affected by saprophytic bacteria in cottonseed homogenates used in the detection of *X. campestris* pv. *malvacearum*.[194] A phage (EP-1) specific for *E. herbicola*, a common saprophyte, was isolated from cottonseed homogenates. Addition of 7×10^7 PFU or more of phage EP-1 to seed homogenates along with *X. campestris* pv. *malvacearum* specific phage XP-1 facilitated the detection of the pathogen of 1.6×10^2 vs. 1.6×10^6 cells per milliliter when only phage XP-1 was present. The modified phage increased the possibility for detecting the pathogen.[194]

The phage-plague method has certain limitations: the phage may become lysogenic and fail to produce plaques, and it may be difficult to detect closely related strains or species of the bacterium if the phage is highly selective. For example, Wallen et al.[195] failed to detect *X. campestris* pv. *phaseoli* var. *fuscans* by this method because the phage did not form plaques, although over 50% of the bean seeds were infected. Also, in bean, cucurbit, tomato, and other seedlots, several bacterial pathogens may be present in a single sample, so that it may be necessary to perform the test using a number of specific phages. The method is not very sensitive. There may be phages which may infect a number of bacteria, and certain phages are strain specific within a species. In such a case, the use of phage for diagnostic purposes is limited.

V. TESTING METHODS FOR SEEDBORNE VIRUSES

The detection of viruses in seeds is more difficult than the detection of bacteria, fungi, or nematodes. Viruses are obligate parasites and differ in their genetic makeup, requiring special techniques for detection. Evidence of their seedborne nature is found by either raising seedlings from infected seeds and observing symptoms, or using assays of seed extracts on indicator plants. Methods have been standardized for routine examination of seed samples using seedling symptomatology (growing-on test) and infectivity tests. In addition, serology is used for quick and specific detection of seedborne viruses. The methods used for the detection of seedborne viruses have been reviewed by Carroll[196] (Figure 45) and Phatak.[197,198]

A. Examination of Dry Seeds
Visual observation of seeds may reveal abnormalities such as discoloration, shriveling, reduced seed size, staining, seed coat necrosis, etc. Some of these abnormalities are correlated with parent plant infection and with seed infection. The association of virus infection in different types of seed abnormalities in certain crop seeds was summarized.[198]

1. Discoloration
Mottling, hilum color break, or hilum bleeding of seeds with light-colored seed coats can indicate that such seeds have come from virus-infected plants. Koshimizu and Iizuka[199] first reported a close relationship between soybean seed coat mottling and SMV and soybean stunt virus (SSV). The pattern of seed coat mottling caused by SSV is in concentric rings and not radial zones or bands characteristic of SMV (Figure 6; see Chapter 1). Later, Kennedy and Cooper[200] found soybean seed coat mottling was the result of a high frequency of natural SMV infection. However, asymptomatic soybean seeds can carry a high percentage of SMV.[197,201] Transmission of SMV is higher

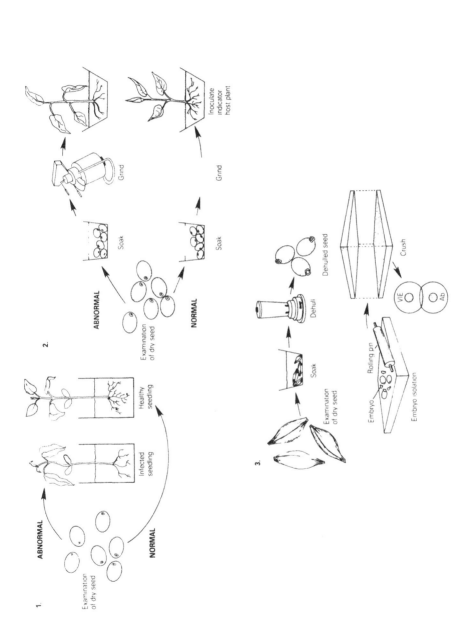

FIGURE 45. Methods for detecting seedborne plant viruses in seeds. (1) Growing-on test; (2) direct seeding test; (3) serological analysis of barley (*Hordeum vulgare*) embryo extracts; (4) latex flocculation method; (5) ELISA test procedure; (6) schematic of the SSEM method.[196]

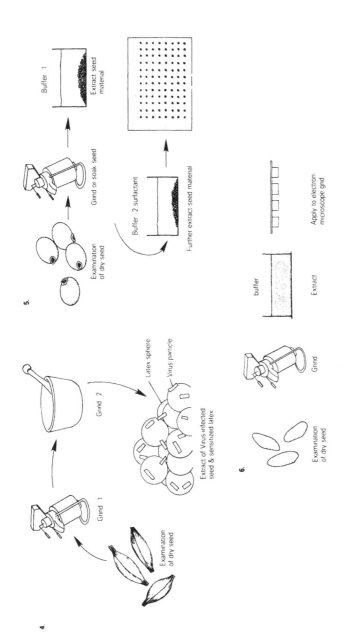

FIGURE 45(4-6)

in mottled than in nonmottled soybean seeds.[202] When seed coat mottling occurs, it indicates that the plant producing the seeds was infected with SMV.[203] Soybean seed coat mottling due to SMV is influenced by host genotype, virus strain, and environment and their interactions with season and location.[204]

The correlation coefficient between seed coat mottling and SMV antigen content was significant for cultivars Beeson, Hodgson, and Steele ($p = 0.01$), and for Amsoy 71 and Williams ($p = 0.05$). The correlation for ten other cultivars was not significant.[205] The percentage of mottled seed from SMV-1-inoculated plants was greater than that from plants inoculated with SMV-2. Mottling was found to be affected by temperature during flowering and at early pod formation. Plants exposed to 20°C during this growth period had a high level of seed coat mottling on two genotypes, whereas exposure to 30°C significantly reduced seed mottling of the susceptible cultivars and virtually eliminated this symptom in moderately susceptible cultivars.[204]

Seed mottling is due to anthocyanin formation in response to virus infection at 20°C. Pigment development in soybean seed coats is due to the accumulation of flavoid compounds such as anthocyanins,[206] which is influenced by temperature. Thus, mottling may not be induced always by virus infection. The extent of mottling and seed transmission depends on the cultivar.[207] The relationship between genotypes of SMV-infected plants and the extent of mottling also was studied. The SMV-infected soybean plants with the genotype ITrw (yellow hilum seeds) produce 78% brown-mottled seeds; with the ItRw genotype (yellow hilum seeds), 38% buff-mottled seeds; and with itrW (buff hilum seeds), 4% buff-mottled seeds.[208]

Seed coat mottling is an indication of neither virus infection of mother plants nor presence of infectious virus in the seeds. In 1967, a correlation was obtained between percent SMV-infected plants and mottled seed, but no correlation was found in 1977.[209] Soybean seed coat mottling, in susceptible soybean cultivars, correlates with temperature during a certain growth period, the virus strain, cultivar, season, location, and other factors.

In peanut, peanut mottle virus (PMV) is transmitted to a higher degree in small and discolored seeds than in average-sized seeds in the cultivar Argentine. The virus causes brown flecking of the seed coat.[210] TMV causes necrosis and deformation of tomato seeds.[211]

2. Reduced Seed Size

Small seeds have a high level of seed transmission for a few viruses. The squash mosaic virus is transmitted in *Cucumis melo* and *C. pepo* by light, poorly filled, and deformed seeds at a higher percentage than by larger, well-filled seeds.[212] BSMV-infected barley seeds tend to be small. About 2% of large barley seeds produced BSMV-affected seedlings, while transmission in small seeds was as high as 81.5%.[213] In some cases, seed transmission may not be influenced by seed size. Light and heavyweight seeds from lettuce mosaic virus (LMV)-infected lettuce plants show equal amounts of seed transmission of LMV.[214]

3. Shriveled and Wrinkled Seeds

Higher levels of transmission of certain viruses was found in shriveled and wrinkled seeds. A high percent (52%) of cowpea aphid-borne mosaic virus in cowpea seeds was recorded in small, shriveled, and intermediate (25%) wrinkled seeds, while smooth seeds did not contain the virus.[215] Transmission of BCMV in Swiss Blanc bean seeds was 23.9 and 43% in smooth and wrinkled seeds, respectively.[216]

The detection of seedborne viruses based on symptoms is neither quantitative nor specific. Symptomless seeds may carry virus infection. Seed abnormalities also are caused by other pathogens. Therefore, it is necessary to confirm virus infection by more accurate and reliable techniques.

B. Biological Tests

Two biological tests, growing-on (seedling symptomatology test) and indicator-inoculation (infectivity test) are used to detect seedborne viruses.

1. Growing-On Test

The growing-on test (Figure 45) is used to determine virus seed transmission. Seeds are planted in blotter paper, sand, soil, vermiculite, or other growth media. Seedlings are examined regularly for virus symptoms, which are influenced by the environment. The test should be conducted under insect-proof conditions to avoid secondary infections, and light and temperature conditions should provide for optimum plant growth and symptom expression.

Incubation conditions for growing-on tests were studied extensively for detection of BCMV in urdbean;[217] BSMV in barley,[197,218-220] LMV in lettuce,[197,221,222] broadbean mottle in white sweet clover, cowpea aphid-borne mosaic in cowpea, SMV in soybean, and squash mosaic virus in muskmelon.[197]

A two-step system was developed to test for BSMV in barley seeds.[220] Initial seeding is done in pots with barley seeds spaced 4 to 5 cm apart in a circle or row. Seedlings showing no chlorotic markings at 26.7°C are retested in the same manner before concluding they are virus free. Later, Hampton et al.[219] reported a light intensity of 107,689 lx at 21 to 30°C ideal for BSMV symptom expression in barley seedlings. Symptoms are ill defined at lower light intensities, which may result in lower counts.

Rohloff[221,222] described a technique for detection of LMV. Lettuce seeds are sown 2.5 cm apart and incubated in the dark for 5 days at 5 to 10°C followed by continuous light for 13 to 19 days at 20°C. Fluorescent lamps with emissions of blue (400 to 510 nm) and red light (600 to 700 nm) are desirable. The first three true leaves are observed for mosaic symptoms by holding the seedlings against diffuse or shaded light. BCMV is detected in urdbean seedlings by growing seeds for 20 days at 20 to 30°C. Symptoms are observed up to the second trifoliolate leaf. Symptoms on primary leaves appear as a downward cupping of leaf margins and reduced leaf size. In subsequent leaves, irregular light and green patches develop. Leaf size is reduced and later the leaves show slight puckering and blistering. When the test was conducted under intense light at 20.8 to 39.0°C, development of systemic symptoms was masked, and with diffuse light at 20 to 30°C, 10% seed transmission was recorded.[217] Primary leaves of soybean seedlings from SMV-infected seeds, after about 10 days, are misshapen with margins curved downward, Typical virus symptoms appear on the first and subsequent trifoliolate leaves at 22 to 24°C.[207]

The blotter method is used for detection of arabis mosaic virus, raspberry ringspot virus, tobacco ringspot virus, and tomato blackring virus in petunia, BSMV in barley.[197] For the first four viruses, seeds are plated on two layers of white blotters in culture plates, incubated for the first 4 days in the dark at about 20°C, then transferred to white fluorescent light with alternating 12-hr light and dark for 10 days at 20°C. Seedlings are examined for symptoms. For BSMV, barley seeds are placed on blotter sheets soaked in water and covered with transparent covers to control moisture loss until after shoot emergence. Incubation is in the dark or subdued light for 2 to 3 days at 24 to 28°C followed by a bright illumination (about 645,834 lx). Tap water is added regularly. Symptoms are observed on the leaves. This method is as effective as the soil test for detection of BSMV.[197]

The reliability of the growing-on test depends upon symptom expression. The test is not reliable when distinct symptoms do not develop; latent infection can occur. The test is practical when characteristic symptoms appear at early stages of seedling development and environmental conditions for symptom development are known. This test is used for testing the presence of BSMV in barley seeds for quarantine purposes in Kansas and Montana[219] and in the Salinas Valley, California for screening commercial lettuce seed lots for LMV.[196,223]

2. Infectivity or Indicator-Inoculation Test

Viruses can be detected in seeds by assaying the extracts of different parts of seeds and seedlings raised from infected seeds on suitable indicator plants (Figure 45). Susceptible hosts which produce local lesions or systemic symptoms are used as indicator plants. This test has been used to detect BCMV in bean and urdbean,[217,224] LMV in lettuce,[225,226] TMV in tomato,[227] and tobacco ringspot virus (TRSV) in soybean.[228]

The BCMV is detected in individual bean seeds by the disc method. The sap from the embryo and cotyledons of individual seeds, after 3 to 4 days of germination, is inoculated onto the surface of the primary leaf of the susceptible bean cultivar, Top Crop. The inoculated leaf is placed on damp filter paper in a closed culture plate under light at 30 to 32°C. In 2 to 3 days, discrete dark to reddish-brown lesions appear.[229,230] The BCMV is detected in urdbean seed coats, cotyledons, and embryos by bioassaying seed extracts onto healthy urdbean seedlings.[217] The LMV can be detected in lettuce seeds using *Chenopodium quinoa* as an indicator plant. The test can detect 1 infected seed out of 700.[223] Grind 10 g seed (about 700 seeds) in 5 mℓ of 0.03 $MN_2HPO_4 \cdot 12H_2O$ buffer at pH 7 with 0.2% sodium diethyldithiocarbamate and 0.5% $NaHSO_3$. To this add 375 mg activated charcoal and inoculate the paste within 10 min onto *C. quinoa* at the 4- to 6-leaf stage. Inoculated plants are incubated in 16-hr light/dark (9000 lx or more) for 5 to 14 days at 20 to 25°C. Symptoms appear as chlorotic lesions and systemic mosaic spots on top leaves. The *C. quinoa* infectivity test for detecting LMV is better than controlled environmental growing-on tests.[197]

Virus indicator plants such as *Chenopodium* spp. are used for cowpea aphidborne mosaic and cowpea ringspot virus in cowpea, and mungbean mosaic and cucumber mosaic in mungbeans.[197] *C. amaranticolor* is used for identification of pea seedborne mosaic virus in pea,[231] cowpea mosaic virus in cowpea,[232] and strawberry latent ringspot virus in celery.[233] Cucumber is used for necrotic ringspot in Montmorency cherry seeds,[234] *Nicotiana glutinosa* and *Phaseolus vulgaris* "Scottia" for TMV in tomato; and *P. vulgaris* "Bountiful" for alfalfa mosaic virus in alfalfa.[235]

It is difficult to estimate seed infection by the indicator-inoculation test unless individual seed extracts are assayed. The number of seeds for an extract preparation depends upon the virus transmission rate. If transmission is low, then more seeds are required. The test is useful for quarantine purposes, where it may be important to know if the virus is present in a seed sample, for highly infectious viruses which produce local lesions or systemic symptoms, for detecting latent virus infection in seedlings not detected by the growing-on test, and for virus identification. The test is used for detection of pea seedborne mosaic virus in pea seed by the Palouse Seed Co. under the supervision of the Idaho and Washington State Departments of Agriculture with *C. amaranticolor* as an indicator plant. One infected seed in ten can be detected.[236]

C. Biochemical Tests

Serology and histochemical staining are used to detect seedborne viruses, with the former being widely used (Figure 45).

1. Serology

Serological methods are for characterization and determining relationships between viruses. The tests are based on the reaction between an antiserum, a blood serum containing specific antibodies produced by injecting laboratory animals with a pure virus preparation, and an antigen-virus protein. The tests are specific since an antibody combines only with the antigen which contains similar groupings of amino-acid sequences. The union of antigen and antibody can be detected in the form of precipitation and agglutination. Details of serological tests for plant-virus identification have been presented.[196] [198,237] Common serological tests used for detection of seedborne viruses are as follows.

a. Microprecipitin Test

The microprecipitin test is a simple serological test in which drops of crude or purified seed extracts combine with their specific antibody and result in a macro- or microscopically visible precipitate. The test can be carried out on microscope slides, in test tubes, or in culture plates. BCMV in urdbean seeds can be detected by this test. A positive reaction is indicated by formation of a precipitate within 1 hr after the adding of BCMV antiserum to embryo extracts of virus-infected seeds.[217] The test is employed in identifying viruses with rod-shaped or isometric particles.

b. Gel-Diffusion Tests

Gel-diffusion tests are performed in semisolid media. Antibodies and antigens diffuse in an agar-gel prior to serological reaction. Rod-shaped viruses and small particles, which tend to aggregate, diffuse poorly and cannot be detected by this method unless the virus particles are cleaved. This method has an advantage over the microprecipitin test in that before the serological reaction takes place, either antibody and antigen or antigen alone diffuses in an electrical field. This test is more reliable and specific than the microprecipitin test.

(1) Single or Radial Diffusion

In single diffusion, the antigen diffuses into the agar containing the antiserum. This test is performed in culture plates or test tubes. Seed or seedling (from infected seeds) extracts are tested by mixing the corresponding antiserum in agar prior to solidification. The virus diffuses radially in culture plates and the procedure is referred to as a "radial diffusion test". The test is used to detect BSMV in barley and wheat seed.[238] This test is better than the double-diffusion test of Hamilton[239] since the former avoids tissue maceration, well cutting, and chemical treatment of an antigen prior to antibody exposure. It involves the use of the primary leaf of seedlings. Embryos are unsuitable for radial diffusion since nonspecific precipitans can occur around the wells.

Agar-gel plates are prepared by mixing diluted D-protein antiserum containing 1% sodium dibutyl-naphthalene sulfonate (Leonil SA detergent) and 0.4% sodium azide with an equal volume of 1% liquid Ionagar No. 2 buffered with 0.1 M tris HCl (pH 7.2) containing 1.7% NaCl. The solution is equilibrated prior to mixing in a 50°C water bath. A 15-mℓ sample of the mixture is poured into a 9-cm plastic culture plate and allowed to solidify in a moist chamber at room temperature. Barley seeds are germinated in a chamber under continuous mist for 5 days at 20°C, and a 1-mm segment of the primary leaf is excised and embedded into the solidified agar. The plates are incubated in a moist chamber for 24 to 36 hr. Any reaction is observed under a binocular microscope at × 10 and 20. Positive reactions are recorded by the presence of opaque areas in the agar gel. In mass testing of barley seed lots (500 to 600 tissues), samples normally are spaced 1 mm within rows and 2 mm between rows in 9-cm culture plates. The single-diffusion test can detect as little as 1 μg/mℓ of degraded virus.[236]

(2) Double Diffusion

In double diffusion, both antibody and antigen diffuse toward each other in a gel medium. At the site where they meet, a positive reaction results in a precipitin line. This test is suited for small, isometric (spherical) viruses which diffuse easily through agar, such as arabis mosaic, blackgram mosaic (a strain of broadbean mottle), cowpea ringspot, cucumber mosaic (mungbean strain), squash mosaic, tobacco ringspot, and tomato blackring.[197] The test is applicable for nonisometric viruses if they are put through a process for viral cleavage. Seeds suspected of carrying long viruses can be treated with an agent that breaks up or degrades virus particles. Degrading agents enhance virus-particle diffusion. In an agar gel at pH 9.0 containing both free-ammonia ions and a detergent, BSMV, brome mosaic virus, cowpea mosaic virus, maize dwarf mosaic virus, SBMV, TMV, and wheat streak mosaic virus diffuse rapidly to form clear precipitin lines.[240]

The double-diffusion test is suitable for the detection of BSMV in barley[241,242] and TMV in tomato.[197] Hamilton[239] developed a method for estimating BSMV in individual barley embryos. Dry barley seeds are dehulled and the embryos separated from the endosperm. A filter-paper disk is placed in each of 100 holes 1-mm deep and 9 mm in diameter in a Lucite® embryo crusher and individual embryos are transferred to each disk. The disc then is sprayed with phosphate buffered saline (PBS) and a top plate with 100 pegs 2-mm long and 8 mm in diameter is pressed into the holes, crushing the embryos. The fluid from the embryos is absorbed by the paper disks. The discs with crushed embryos are transferred to serological test plates. Embryos showing immuno-precipitates are recorded after 12 to 24 hr at 25°C.

The embryo test was modified so that antisera are elicited using sodium dodecyl sulfate (SDS)-treated preparations or partially purified BSMV and the agar gel is amended with SDS. The SDS disrupts virus particles in the agar medium.[196] The test is used by the Montana Seed Testing Laboratory. Two hundred embryos from each seed lot are tested. Filter-paper disks serve as sero-reactant depots.[236,242] Blackeye mosaic virus in cowpea and SMV in soybean have been detected in hypocotyls of 4- to 5-day-old seedlings using the SDS method. Antigens consist either of hypocotyl extracts in SDS or hypocotyl disks embedded in agar containing SDS. The percentage of infected cowpea and soybean seeds recorded was similar to the percentage infection determined by seedling symptoms.[243]

In detecting pea seedborne mosaic virus in pea, crystals may form around some antigen wells. Precipitin band formation is clear and separation from nonspecific precipitate rings around the antiserum well is more distinct on Bacto agar than on special Noble agar. A reduction in SDS concentration in Bacto agar from 0.5 to 0.375% and an increased concentration of sodium azide (NaN_3) from 1.0 to 1.5% gives a more distinct separation. This is an improvement over the diffusion method suggested by Hamilton and Nichols[244] for detection of pea seedborne mosaic virus.[245]

Nonspecific reactions may occur with legume seed extracts in gel-diffusion tests. Cockbain et al.[246] reported that *Vicia faba* embryos cause the agar to cloud, interfering with formation of virus-specific precipitin lines. Lister[247] found nonspecific precipitin formation when detecting TRSV in soybean using the double-diffusion test. Phatak[197] assumed that a high content of lectin (hemagglutins) in soybean seeds was responsible for unsatisfactory results using the hemagglutination test for SMV in seed extracts. Lectins occur at high concentrations in certain legume seeds.[248] Double diffusion can detect virus concentrations of 10 to 25 µg/mℓ.[196]

c. Agglutination Tests

In this method, materials such as bentonite, chloroplasts, RBC of sheep, or synthetic polystyrene latex are used as reaction sites. After the reaction takes place, the carrier particles exhibit agglutination or clumping. These tests are suitable for elongate virus particles which do not diffuse through a gel medium.[197]

Chloroplast agglutination — Jermoljev and Chod[249] used this technique for detection of BCMV in bean seeds. Drops of germinated seed extract, used as antigen, are placed over paper disks soaked in BCMV antiserum. A positive serological reaction is indicated by the presence of chloroplast clumping. The test is carried out with crude noncentrifuged seedling extracts. Chloroplasts and other cell particles also get enmeshed in the network of the precipitate. Hence, there are more chances of nonspecific precipitation and, thus, the test is less sensitive than others.

Latex agglutination test — To avoid nonspecific precipitation when using biologically active material, a biologically inert material, such as polystyrene latex balls, can be used (Figure 45).[250] The use of this test for early, rapid diagnosis of seedborne viruses is advocated by Phatak.[197] It is specific for diagnosis of BSMV and SMV in young seedlings and in barley and soybean seeds. Symptomless infections have been detected.[197] A technique for conducting the latex test in capillary tubes for BSMV in barley was introduced by Marcussen and Lundsgaard.[251] Barley seeds are incubated for 7 days on wet blotters, the shoot apexes ground in buffer solution, and the sap suspension mixed with anti-BSMV sensitized latex in capillary tubes. Aggregation of the latex spheres following agitation of the suspension indicates that the samples are infected with BSMV. This method is recommended for routine seed health testing laboratories.[252]

d. Labeled Antibodies

The use of labeled antibodies antiserum has been used to detect seedborne viruses in individual seeds with accuracy and specificity.

e. Immunofluorescence Microscopy Test

This test is used for detecting SMV in soybean seeds. Thin sections of plumules from germinating seeds are cut and treated with a specific antiserum and stained with anti-rabbit-sheep serum conjugated with fluochrome-fluorescein isothiocyanate. The portions then are washed for 30 min and examined under a fluorescence microscope. Blue fluorescence is obtained from infected plumules.[197]

f. Radioisotope-Labeled Antibody

Antiserum containing radiolabeled antibodies is used for detecting squash mosaic virus in cantaloupe seeds. The seeds are cut into pieces and inoculated with labeled antiserum. Radioactivity is measured after washing the seed pieces. The presence of radioactivity indicates positive detection of seedborne infection.[253]

g. ELISA

The ELISA test (Figure 45), also known as the double antibody sandwich procedure, is used for detecting animal and plant viruses.[254,255] The test is sensitive to arabis mosaic virus and plum pox virus, enabling assays at virus concentrations as low as 10 to 100 $\mu g/m\ell$ in purified and crude plant extracts.[255] Conventional serological techniques cannot be used for many viruses because of low concentration, unsuitable particle morphology, or the presence of virus inhibitors or inactivators in plant extracts. These can largely be overcome by using ELISA.[254]

Bossennec and Maury[256] first used ELISA to detect SMV in soybean seeds. The test has been used for detecting BSMV in barley,[257] LMV in lettuce,[257,258] prune dwarf virus in *Prunus avium*,[259] SMV[205,247] in soybean, and BCMV in bean,[258] and TRSV in soybean.[247,257] The ELISA is the most sensitive of all plant virus serological tests. Viruses were detectable in extracts from seed samples with less than 1% TRSV and 2 to 4% SMV. The use of soybean seedlings for testing for SMV is more sensitive than testing seeds.[247] In bean seeds, BCMV can be detected in 1 of 2000 embryos and LMV in 1 of 1400 lettuce seeds.[258]

The method involves adsorption of specific antiserums onto a polystyrene plate. Excess antibody molecules are washed from the plates with water. A seed extract is added and the excess washed off. The enzyme-labeled specific antibody (usually alkaline phosphates) is added, which results in the conjugation of a double antibody antigen. Excess labeled antibody molecules are removed by washing, and the enzyme substrate (usually paranitrophenyl phosphate) is added. The reaction product is observed visually or quantitatively recorded with a spectrophotometer at 405 nm. The presence of virus is indicated by a yellow color. Virus concentrations as low as 0.1 $\mu g/m\ell$ can be detected.[196,205,247]

The ELISA test is suitable for both filamentous and isometric viruses, and may be applicable to large-seeded species, such as cereals and legumes.[247,254] A limiting factor may be that the test is relatively complex and expensive.

h. Serologically Specific Electron Microscopy (SSEM)

This method (Figure 45) was first used by Derrick and Brlansky[260] to detect seedborne viruses. The following viruses have been detected using this method: BCMV in bean, LMV in lettuce, TRSV in soybean,[261] pea seedborne mosaic virus in pea,[244] and SMV in soybean.[243,261]

First coat a copper grid (200 mesh) with parlodian film and float it on virus antiserum diluted to 1:100 to 1:5000 in tris buffer for 30 min. Wash the grid to remove unadsorbed serum proteins and then float it on a crude seed extract in buffer for 10 min to 24 hr. Wash the grid again and later stain it with 5% uranyl acetate for about 2 min. Wash it with distilled water or 95% ethanol, dry, and examine it under a transmission electron microscope.[196,260,261]

Virus particles of TRSV in soybean, BSMV in barley and SMV in soybean are detected in 1:100 parts of seed extract, and LMV is 1:100.[261] The SSEM is superior to ELISA for detection of pea seedborne mosaic virus in pea seeds and for detecting the virus in seed lots containing 1 to 5% infected seeds, with none detected using ELISA. The superiority of SSEM may be due to the high A_{405} (0.1 to 1.5) of healthy seed homogenates. Such a background interferes with detecting very low levels of virus-infected seeds by ELISA. Antibodies to normal pea proteins might be bound on the plastic surface and would compete for the enzyme-labeled antibody in proportion to their concentration. The use of antiserum cross-absorbed with normal pea proteins might alleviate this difficulty. However, both ELISA and SSEM are uniquely advantageous in screening seed lots before introduction into a germ pool.[244]

i. Solid-Phase Radioimmunoassay

This test is similar to ELISA and is equally effective in detecting SMV in soybean seeds. The anti-SMV immunoglobulin G (IgG) bodies are adsorbed to polystyrene balls. Detection is achieved by adding antibody conjugated with radioactive tritium (^3H) and counting the bound antibody-virus conjugated antibody sandwich in a liquid scintillation counter. The method is highly quantitative, with determination of virus quantity in a sample determined by reference to a standard curve run concurrently with the assay.[205]

2. Staining Techniques

A number of stains such as acridine orange, giemsa, phloxin, trypan blue, etc. can detect virus infections in plant tissues. The use of acridine orange "fluorescent stain" can detect cowpea mosaic virus in cowpea seeds.[232] Cowpea seeds are soaked in water for 2 days, then sections are cut from the germinating embryos, stained with acridine orange, and viewed under a fluorescent microscope. The virus is detected by aggregates of red fluorescing material in infected tissues. The presence of the virus is confirmed by indexing germinating embryos on an indicator plant, C. amaranticolor.

D. Biophysical Tests

Electron microscopy is used to study biophysical properties of plant viruses. Viruses can be detected in seed extracts or ultrathin sections of infected seed parts or seedlings.

Ultrathin sectioning, cut-squeeze, squash homogenate, or dip methods are used to prepare tissues for detection of seedborne viruses using electron microscopy. Large-scale use is not prescribed. Negative results should not be used to show the absence of virus since virus concentration may be too low to detect. BSMV can be found in ultrathin sections of the embryo and endosperm of barley and wheat seeds which have as high a concentration of virus particles as leaves. Individual green seeds can be screened for BSMV infection using electron microscopy.[262] The BSMV was found in pollen and young seedling leaves from infected seeds.[263,264] Cherry leaf roll virus and strawberry latent ringspot virus were detected in tubular inclusion bodies in seed homogenates from *N. rustica* and celery, respectively, using the squash homogenate method.[265] BCMV was found in bean ovules using the dip method.[266] Particles of BCMV, SMV, and BSMV were found in plumules, extracted after a 2-hr pre-soak (12 hr for barley) in distilled water, of bean, soybean, and barley seeds, respectively, using the squash homogenate method. Individual plumules are squashed in drops of negative stain (2% ammonium molybdate or 2% phosphotungstic acid, pH 7.0) on glass slides. The homogenate then is transferred with a capillary tube to formvar-coated carbon-reinforced copper grids for examination in an electron microscope. Virus particles resembling BCMV, BSMV, and SMV were found using this test.[197] LMV is not detected by this method, probably due to low concentrations of the virus in seeds.[197,267] Cytoplasmic inclusions (cylindrical) were detected in blackeye mosaic virus-infected cowpea hypocotyls and SMV-infected soybean hypocotyls by light microscopy.[243]

E. Contact Radiography with X-Rays

Contact radiography with X-rays is used to determine insect damage to seeds. The technique was suggested for detection of BSMV in barley seeds and runner bean mosaic virus in runner bean seeds by Phatak and Summanwar.[213] Opacity in barley is reduced in BSMV-infected seeds, most of which produce infected seedlings. Similarly, this differences in opacity was noticed in healthy and infected runner bean seeds. However, the technique has not received much attention.

VI. TESTING METHODS FOR SEEDBORNE NEMATODES

Nematodes associated with seeds can be detected by visual observation or soaking seeds in water. Nematode identification, based on body structure and anatomy, should be done with the help of manuals or descriptions published by the Commonwealth Institute of Helminthology, Kew, England, or the pictorial keys of Mai and Lyon.[268]

A. Examination of Dry Seeds

The seed-gall nematode, *Anguina* spp., which replaces grains with gall-like structures, can be detected by visual examination. The galls or cockles are small, dark purplish-black, or dull. Diagnosis is confirmed by soaking galls in water for 1 hr and cutting them into pieces in drops of water. Active larvae are released into the water and can be seen under a binocular microscope. *Anguina agropyronifloris*, *A. agrostis*, and *A. tritici* also can be detected.

B. Examination of Water on Soaking Seeds

Nematodes passively associated with seeds, such as *Aphelenchoides besseyi*, *Ditylenchus angustus*, and *D. dipsaci*, are released from seeds on soaking in water. Ou[269] recommended using warm water for release of *A. besseyi* from rice seeds; dehulled seeds are soaked for 12 to 24 hr. Some nematodes move out in a few hours but a majority emerge after 12 to 24 hr. For a recovery of the nematodes up to 89%, soak rice seeds and place them over a nylon mesh suspended in water for 12 to 15 hr at 28°C.[270] Soaking seeds assures that moisture reaches dormant nematodes, and a quicker release of nematode larvae.

D. angustus and *D. dipsaci* are released from rice and onion seeds, respectively, by water soaking.[271-273] For detection of *D. dipsaci* in clover and onion seeds, place seeds on filter paper in a hopper closed with a tap at the bottom. The hopper is sprayed with a water mist, releasing larvae which sink through the filter paper and into the water just above the tap. Drops of this water are examined for larvae. Cysts of *Heterodera* spp. can be recovered by flotation. Cysts float on top which are recovered on a sieve with 0.25-mm openings.[274] Water temperature influences the recovery of *A. besseyi* from rice seeds. Larvae are released at 20°C but not at 35°C.[275]

C. Fenwick's Small Culture Cells

Fenwick culture cells are used to detect *D. dipsaci* in onion seeds. The small, circular glass cups (culture cells) are about 1 cm in diameter and 0.4-cm deep. A battery of 50, arranged in rows, is cemented to a sheet of glass which is cut to fit 15-cm culture plates.[276] For each seed sample, 20 dishes of 50 cells each are set up with a single onion seed in each cup containing a small quantity of sterile distilled water. Each cell is examined for larvae the next day.[277]

D. Extraction of Nematodes from Plant Pieces and Soil Clods

Nematodes associated with seed lots in contaminated soil clods or peds and infected plant pieces can be detected by the methods of Ayoub,[278] Dropkin,[274] and Southey.[279]

REFERENCES

1. International Seed Testing Association, Rules for seed testing, *Seed Sci. Technol.*, 4, 3, 1976.
2. Agarwal, V. K., Assessment of seed-borne infection and treatment of wheat seeds for the control of loose smut infection, *Seed Sci. Technol.*, 9, 725, 1980.
3. Jørgensen, J., The Doyer filter paper method in health testing of barley seed, *Proc. Int. Seed Test. Assoc.*, 36, 325, 1971.
4. Agarwal, V. K., Techniques for the detection of seed-borne fungi, *Seed. Res.*, 4, 24, 1976.
5. Naumova, N. A., *Testing of Seeds for Fungous and Bacterial Infections*, (transl. from Russian), 3rd ed., Israel Program for Scientific Translations, Jerusalem, 1972, 145.
6. Neergaard, P., Detection of seed-borne pathogens by culture tests, *Seed Sci. Technol.*, 1, 217, 1973.
7. de Tempe, J., Routine methods for determining the health condition of seeds in the seed testing station, *Proc. Int. Seed Test. Assoc.*, 35, 257, 1970.
8. Ilyas, M. B., Dhingra, O. D., Ellis, M. A., and Sinclair, J. B., Location of mycelium of *Diaporthe phaseolorum* var. *sojae* and *Cercospora kikuchii* in infected soybean seeds, *Plant Dis. Rep.*, 59, 17, 1975.
9. Johnson, H. W. and Jones, J. P., Purple stain of guar, *Phytopathology*, 52, 269, 1962.
10. Sherwin, H. S. and Kreitlow, K. W., Discoloration of soybean seeds by the frogeye fungus, *Cercospora sojina*, *Phytopathology*, 42, 568, 1952.
11. Gangopadhyay, S., Wyllie, T. D., and Luedders, V. D., Charcoal rot diseases of soybean transmitted by seeds, *Plant Dis. Rep.*, 54, 1088, 1970.
12. Hepperly, P. R. and Sinclair, J. B., Quality losses in *Phomopsis* infected soybean seeds, *Phytopathology*, 68, 1684, 1978.
13. Rodriguez-Marcano, A. and Sinclair, J. B., Fruiting structures of *Colletotrichum dematium* var. *truncata* and *Phomopsis sojae* formed in soybean seeds, *Plant Dis. Rep.*, 62, 973, 1978.

14. Schneider, R. W., Dhingra, O. D., Nicholson, J. F., and Sinclair, J. B., *Colletotrichum truncatum* borne within the seed coat of soybean, *Phytopathology*, 64, 154, 1974.
15. Agarwal, V. K., Seed-borne Fungi of Rice, Wheat, Blackgram, Greengram and Soybean Grown at G. B. Pant University Agricultural Technical Farm, Pantnagar, India, Research Monograph, Institute of Seed Pathology for Developing Countries, Copenhagen, 1970, 40.
16. Dharam Vir, Adlakha, K. L., Joshi, L. M., and Pathak, K. D., Preliminary note on the occurrence of black-point disease of wheat in India, *Indian Phytopathol.*, 21, 234, 1968.
17. Hewett, P. D., Disease testing in a seed improvement programme, in *Seed Pathology — Problems and Progress*, Yorinori, J. T., Sinclair, J. B., Mehta, Y. R., and Mohan, S. K., Eds., Fundação Instituto Agronômico do Paraná, IAPAR, Londrina, Brazil, 1979, 72.
18. Prabhu, A. S. and Prasada, R., Investigations on the leaf blight disease of wheat caused by *Alternaria triticina*, in *Plant Disease Problems*, Raychaudhuri, S. P., Ed., Indian Phytopathological Society, Indian Agriculture Research Institute, New Delhi, 1970, 17.
19. Adlakha, K. L. and Joshi, L. M., Black point of wheat, *Indian Phytopathol.*, 27, 41, 1974.
20. Agarwal, V. K., Mathur, S. B., and Neergaard, P., Some aspects of seed health testing with respect to seedborne fungi of rice, wheat blackgram, greengram and soybean grown in India, *Indian Phytopathol.*, 25, 91, 1972.
21. Schroeder, H. W., Grain discoloration in Belle Patna rice, *Plant Dis. Rep.*, 48, 288, 1964.
22. Wright, W. R. and Billeter, B. A., "Red kernel" disease of sweetcorn on the retail market, *Plant Dis. Rep.*, 58, 1065, 1974.
23. Sherf, A. F., *Septoria avenae* Frank on *Avena sativa* L., in *Handbook on Seed Testing*, Association of Official Seed Analysts, Junction City, Ore., 1958, 32.
24. Dhanraj, K. S. and Mathur, S. B., Ear rot of maize caused by *Cephalosporium acremonium* Corda, a new record for India, *Indian Phytopathol.*, 18, 393, 1965.
25. Lal, S., Raju, C. A., and Agarwal, V. K., A note on the fungi associated with white streaks on maize kernels, *Seed Res.*, 5, 64, 1977.
26. Ram Nath, Mathur, S. B., and Neergaard, P., Seed-borne fungi of mungbean (*Phaseolus aureus* Roxb.) from India and their significance, *Proc. Int. Seed Test. Assoc.*, 35, 225, 1970.
27. Steadman, J. R., White mold disease initiation, development and control in Nebraska, in *Rep. Bean Improvement Cooperative and Natl. Dry Bean Res. Assoc. Conf.*, Rochester, N.Y., 1975, 2.
28. Basuchaudhari, K. C. and Pal, A. K., Infection of sunnhemp seeds with *Fusarium* spp., *Seed Sci. Technol.*, 9, 729, 1981.
29. Moravcik, E., Effect of Ascochyta blight of pea seeds on the germination and emergence, in *Proc. 19th Int. Seed Testing Assoc. Congr.*, Vienna, 1980.
30. Wenzl, H., Studies on the infection of beet seeds by *Cercospora beticola* in relation to weather and climate, *Pflanzenschatzberichte*, 23, 116, 1959.
31. Ranganathaiah, K. G. and Mathur, S. B., Seed health testing for *Eleusine coracana* with special reference to *Drechslera nodulosa* and *Pyricularia grisea*, *Seed Sci. Technol.*, 6, 943, 1978.
32. Plurad, S. B. and Daugherty, D. M., Growth parameters of *Nematospora coryli* Peglion (Ascomycetaceae) and progress of infection in inoculated soybeans, *Trans. Mo. Acad. Sci.*, 4, 125, 1970.
33. Preston, D. A. and Ray, W. W., Yeast spot disease of soybean reported from Oklahoma and North Carolina, *Plant Dis. Rep.*, 27, 601, 1943.
34. Halfon-Meiri, A., *Alternaria crassa*, a seed-borne fungus of *Datura*, *Plant Dis. Rep.*, 54, 442, 1970.
35. Kernkamp, M. F. and Hemerick, G. A., Alfalfa seed losses due to *Ascochyta imperfecta*, *Phytopathology*, 42, 468, 1952.
36. Miller, C. S. and Colhoun, J., Fusarium diseases of cereals. VI. Epidemiology of *Fusarium nivale* on wheat, *Trans. Br. Mycol. Soc.*, 52, 195, 1969.
37. Cooke, B. M. and Jones, D. G., The epidemiology of *Septoria tritici* and *S. nodorum*. II. Comparative studies of head infection by *Septoria tritici* and *S. nodorum* on spring wheat, *Trans. Br. Mycol. Soc.*, 54, 395, 1970.
38. Cunfer, B. M., The incidence of *Septoria nodorum* in wheat seed, *Phytopathology*, 68, 832, 1978.
39. Schiller, C. T. and Sinclair, J. B., Soybean seed lot contamination by *Melanopsichium pennsylvanicum* smut galls, *Phytopathology*, 69, 605, 1979.
40. Langdon, R. F. N. and Cusack, A. N., Soybean smut, *Australas. Plant Pathol.*, 7, 43, 1978.
41. Halfon-Meiri, A., Seed transmission of safflower rust (*Puccinia carthami*) in Israel, *Seed Sci. Technol.*, 11, 835, 1983.
42. Johnson, H. W. and Lefebvre, C. L., Downy mildew on soybean seeds, *Plant Dis. Rep.*, 26, 49, 1942.
43. Pathak, V. K., Mathur, S. B., and Neergaard, P., Detection of *Peronospora manshurica* (Naum.) Syd. in seeds of soybean, *Glycine max*, *EPPO Bull.*, 8, 21, 1978.
44. Jones, F. R. and Torrie, J. H., Systemic infection of downy mildew in soybean and alfalfa, *Phytopathology*, 36, 1057, 1946.

45. Maisurian, N. A. and Mordashov, A. I., Method of separating seed-infected by Fusariosis from soybean sowing material, *Sel. Seed-Gr. (Moscow)*, 28, 77, 1963.
46. Kietreiber, M., An improvement of the Tilletia testing method for wheat by the use of special filter papers, in *Proc. 15th Int. Workshop on Seed Pathology*, Paris, 1975.
47. Pirson, H., A simple method for the detection of bunt and covered smut spores as contaminants on cereal seeds, Proc. 16th Int. Workshop on Seed Pathology, Karlsruhe, W. Germany, 1978, 54.
48. Neergaard, P., Lambat, A. K., and Mathur, S. B., Seed health testing of rice. III. Testing procedures for detection of *Pyricularia oryzae* cav., *Proc. Int. Seed Test. Assoc.*, 35, 157, 1970.
49. Hewett, P. D., Methods for detecting viable seed-borne infection with celery leaf spot, *J. Natl. Inst. Agric. Bot.*, 9, 174, 1962.
50. Shetty, H. S., Khanzada, A. K., Mathur, S. B., and Neergaard, P., Procedure for detecting seed-borne inoculum of *Sclerospora graminicola* (Sacc.) Schroet. in pearl millet (*Pennisetum typhoides* (Burm) Stapf. & Hubb.), *Seed Sci. Technol.*, 6, 943, 1978.
51. Benoit, M. A. and Mathur, S. B., Identification of species of *Curvularia* on rice seed, *Proc. Int. Seed Test. Assoc.*, 35, 99, 1970.
52. Chidambaran, P., Mathur, S. B., and Neergaard, P., Identification of seed-borne *Drechslera* species, *Friesia*, 10, 165, 1973.
53. Kulshrestha, D. D., Mathur, S. B., and Neergaard, P., Identification of seed-borne species of *Colletotrichum*, *Friesia*, 11, 116, 1976.
54. Hagborg, W. A. F., Warner, G. M., and Phillips, N. A., Use of 2-4,D as an inhibitor of germination in routine examination of beans for seedborne infection, *Science*, 111, 91, 1950.
55. Neergaard, P., Arsberetning Vedrende Frpatologisk Kontrol. St. Plantetilsyn, 1 April, 1953—31 May, 1954, *Statens Plantetilsyn (Plant Protection Service)*, Copenhagen, 1956, 17.
56. Neergaard, P. and Saad, A., Seed health testing of rice. A contribution to development of laboratory routine testing methods, *Indian Phytopathol.*, 15, 85, 1962.
57. Maquire, J. D. and Gabrielson, R. L., Factors affecting the sensitivity of 2,4-D assays of crucifer seed for *Phoma lingam*, *Phytopathology*, 69, 1037, 1979.
58. Limonard, T., Ecological aspects of seed health testing, *Proc. Int. Seed Test. Assoc.*, 33, 1, 1968.
59. Singh, D. V., Mathur, S. B., and Neergaard, P., Seed health testing of maize: evaluation of testing techniques, with special reference to *Drechslera maydis*, *Seed Sci. Technol.*, 2, 349, 1974.
60. de Tempe, J. and Limonard, L., Seed-fungal-bacterial interactions, *Seed Sci. Technol.*, 1, 203, 1973.
61. Michail, S. H., Mathur, S. B., and Neergaard, P., Seed health testing for *Rhizoctonia solani* on blotters, *Seed Sci. Technol.*, 5, 603, 1977.
62. Muskett, A. E. and Malone, J. P., The Ulster method for the examination of flax seed for the presence of seedborne parasites, *Ann. Appl. Biol.*, 28, 8, 1941.
63. Malone, J. P., Studies on seed health. IV. The application of heat to seedcoats as an aid in the detection of *Pyrenophora avenae* by the Ulster method, *Proc. Int. Seed Test. Assoc.*, 27, 856, 1962.
64. Miller, J. J., Peers, D. J., and Neal, R. W., A comparison of the effects of several concentrations of oxgall in platings of soil fungi, *Can. J. Bot.*, 29, 26, 1951.
65. Nash, S. M. and Snyder, W. C., Quantitative estimations by plate counts of propagules of the bean root rot *Fusarium* in field soils, *Phytopathology*, 52, 567, 1962.
66. de Tempe, J., Testing cereal seeds for *Fusarium* infection in the Netherlands, *Seed Sci. Technol.*, 1, 845, 1973.
67. Malalasekera, R. A. P. and Colhoun, J., Fusarium diseases of cereals. V. A Technique for the examination of wheat seed infected with *Fusarium culmorum*, *Trans. Br. Mycol. Soc.*, 52, 187, 1969.
68. Agarwal, V. K. and Singh, O. V., Routine testing of crop seeds for *Fusarium moniliforme* with a selective medium, *Seed Res.*, 2, 19, 1974.
69. Mangan, A., A new method for the detection of *Pleospora bjorlingii* infection of sugarbeet seed, *Trans. Br. Mycol. Soc.*, 57, 169, 1971.
70. Mangan, A., Detection of *Pleospora bjorlingii* infection on sugarbeet seed, *Seed Sci. Technol.*, 2, 343, 1974.
71. Bugbee, W. M., A selective medium for the enumeration and isolation of *Phoma betae* from soil and seed, *Phytopathology*, 64, 706, 1974.
72. de Tempe, J., Testing sugarbeet seed for *Phoma betae*, *Seed Sci. Technol.*, 6, 927, 1978.
73. Prabhu, A. S. and Prasada, R., Evaluation of seed infection caused by *Alternaria triticina* in wheat, *Proc. Int. Seed Test. Assoc.*, 32, 647, 1967.
74. Kulik, M. M., Detection of seed-borne microorganisms: other tests, *Seed Sci. Technol.*, 1, 255, 1973.
75. Kulik, M. M., Comparison of blotters and guaiacol agar for detection of *Helminthosporium oryzae* and *Trichoconis padwickii* in rice seeds, *Phytopathology*, 65, 1325, 1975.
76. Kulik, M. M., Detection of *Helminthosporium oryzae* in rice seeds using guaiacol agar, *Proc. Am. Phytopathol. Soc.*, 1, 39, 1974.

77. Kulik, M. M. and Koch, E. J., A referee tests of the blotter and guaiacol agar methods for the detection of *Helminthosporium oryzae* and *Trichoconis padwickii* in rice seeds, *J. Seed Technol.*, 1, 71, 1976.

78. de Tempe, J., Routine methods for determining the health condition of seed in the seed testing station, *Proc. Int. Seed Test. Assoc.*, 26, 27, 1961.

79. Leach, C. M., Environmental conditions and incubation period in seed health testing, in *Seed Pathology — Problems and Progress*, Yorinori, J. T., Sinclair, J. B., Mehta, Y. R., and Mohan, S. K., Eds., Fundação Instituto Agronômico do Paraná, IAPAR, Londrina, Brazil, 1979, 89.

80. Leach, C. M., The light factor in the detection and identification of seed-borne fungi, *Proc. Int. Seed Test. Assoc.*, 32, 565, 1967.

81. Leach, C. M., Sporulation of diverse species of fungi under near-ultra-violet radiation, *Can. J. Bot.*, 40, 151, 1962.

82. Malone, J. P., Studies on seed health. II. The use of malt extract in the routine examination of seeds for health, *Proc. Int. Seed Test. Assoc.*, 24, 179, 1959.

83. Leach, C. M. and Trione, E. J., Action spectra for light induced sporulation of the fungi *Pleospora herbarum* and *Alternaria dauci*, *Photochem. Photobiol.*, 5, 621, 1966.

84. Trione, E. J., Leach, C. M., and Mutch, J., Sporogenic substances isolated from fungi, *Nature (London)*, 212, 163, 1966.

85. Leach, C. M., Interaction of near-ultraviolet light and temperature on sporulation of the fungi *Alternaria, Cercosporella, Fusarium, Helminthosporium* and *Stemphylium*, *Can. J. Bot.*, 45, 1999, 1967.

86. Leach, C. M., A comparison of the light requirements necessary to induce reproduction in three fungi, *Trans. Br. Mycol. Soc.*, 46, 302, 1963.

87. de Tempe, J., The quantitative effect of light and moisture on carrot seed infections in blotter medium, *Proc. Int. Seed Test. Assoc.*, 33, 547, 1968.

88. Mathur, S. B. and Neergaard, P., Seed health testing of rice. IV. Effect of light and temperature on seed-borne fungi in the blotter test, *Proc. Int. Seed Test. Assoc.*, 37, 723, 1972.

89. Aulakh, K. S., Mathur, S. B., and Neergaard, P., Seed health testing of rice and comparison of field incidence with laboratory counts of *Drechslera oryzae* and *Pyricularia oryzae*, *Seed Sci. Technol.*, 2, 393, 1974.

90. Kang, C. S., Neergaard, P., and Mathur, S. B., Seed health testing of rice. VI. Detection of seed-borne fungi on blotters under different incubation conditions of light and temperature, *Proc. Int. Seed Test. Assoc.*, 37, 731, 1972.

91. Leach, C. M., Protect the eyes from NUV, *Seed Pathol. News*, 11, 4, 1978.

92. Kolk, H. and Karlberg, S., Studies on the blotter method for determination of seedling diseases in cereals, Preprint 27, *Proc. 16th Int. Seed Testing Assoc. Congr.*, Washington, D.C., 1971, 12.

93. Grybauskas, A. P., Sinclair, J. B., and Foor, S. R., Surface disinfection of soybean seeds for selective recovery of seedborne microorganisms, *Plant Dis. Rep.*, 63, 887, 1979.

94. Hampton, J. G. and Matthews, D., The evaluation of seed-borne *Drechslera teres* in barley — a note on methodology, *Seed Sci. Technol.*, 8, 371, 1980.

95. Hewett, P. D., Pretreatment in seed health testing. II. Duration of hypochlorite pretreatment in the agar plate test for *Ascochyta* spp., *Seed Sci. Technol.*, 7, 83, 1979.

96. Yorinori, J. T. and Sinclair, J. B., Harvest and assay methods for seedborne fungi in soybeans and their pathogenicity, *Trop. Plant Dis.*, 1, 53, 1983.

97. Mathur, S. B. and Ram Nath, *Pestalotia guepini* Desm. in seeds of *Sorghum vulgare* Pres., *Proc. Int. Seed Test. Assoc.*, 35, 165, 1970.

98. Derbyshire, D. M., A study of seed-borne infection of tomato by *Didymella lycopersici* Kleb., *Proc. Int. Seed Test. Assoc.*, 26, 61, 1961.

99. Ram Nath, Neergaard, P., and Mathur, S. B., Identification of *Fusarium* species on seeds as they occur in blotter test, *Proc. Int. Seed Test. Assoc.*, 35, 121, 1970.

100. Mew, I. C. and Kommedahl, T., Interaction among microorganisms occurring naturally and applied to pericarps of corn kernels, *Plant Dis. Rep.*, 56, 861, 1972.

101. Marshall, G. M., Germination and seedling evaluation of peas infected with *Ascochyta* spp., *J. Natl. Inst. Agric. Bot.*, 9, 170, 1962.

102. de Tempe, J., Health testing of flax seed, *Proc. Int. Seed Test. Assoc.*, 28, 3, 1963.

103. Hewett, P. D., Testing carrot seed infected with *Alternaria porri* f. sp. *dauci*, *Proc. Int. Seed Test. Assoc.*, 31, 1, 1964.

104. Halfon-Meiri, A. and Kulik, M. M., *Septoria nodorum* infection of wheat seeds produced in Pennsylvania, *Plant Dis. Rep.*, 61, 867, 1977.

105. Guerrero, F. G., Mathur, S. B., and Neergaard, P., Seed health testing of rice. V. Seed-borne fungi associated with abnormal seedlings of rice, *Proc. Int. Seed Test. Assoc.*, 37, 985, 1972.

106. Doyer, L. C., *Manual for the Determination of Seed-borne Diseases*, International Seed Testing Association, Wageningen, The Netherlands, 1938, 59.

107. Fakir, G. A., Rao, M. H., and Thirumalachar, M. J., Seed transmission of *Macrophomina phaseolina* in sunflower, *Plant Dis. Rep.*, 60, 736, 1976.
108. McDonald, W. C. and Martens, J. W., Leaf and stem spot of sunflowers caused by *Alternaria zinniae*, *Phytopathology*, 53, 93, 1963.
109. Jørgensen, J., Occurrence and importance of seed-borne inoculum of *Cochliobolus sativus* on barley seed in Denmark, *Acta Agric. Scand.*, 24, 49, 1972.
110. Pirson, H., *Phoma lingam* (Tode Ex Schw.) Desm, on *Brassica* spp.: an attempt to simplify the evaluation on seeds, in *Proc. 16th Int. Workshop on Seed Pathology*, Karlsruhe, W. Germany, 1978, 53.
111. Jørgensen, J., Comparative testing of barley seed for inoculum of *Pyrenophora graminea* and *P. teres* in greenhouse and field, *Seed Sci. Technol.*, 8, 377, 1980.
112. Khare, M. N., Mathur, S. B., and Neergaard, P., A seedling symptom test for detection of *Septoria nodorum* in wheat seed, *Seed Sci. Technol.*, 5, 613, 1977.
113. Hiltner, L., Technische vorschriften für die Prüfung von saatgüt, gültig vom I. Juli 1916 an B. Besonderer Teil I. Getreide, *Landwirtsch. Vers. Stn*, 89, 379, 1917.
114. Halfon-Meiri, A. and Volcani, Z., A combined method for detecting *Colletotrichum gossypii* and *Xanthomonas malavacearum* in cotton seed, *Seed Sci. Technol.*, 5, 129, 1971.
115. Kommedahl, T., Lang, D. S., and Blanchard, K. L., Detection of kernels infected with *Helminthosporium maydis* in commercial samples of corn produced in 1970, *Plant Dis. Rep.*, 55, 726, 1971.
116. Jorgensen, J. and Klitogard, K., Blackening of seedlings of broadbean (*Vicia faba*) during germination in sand, *Saertryk Statsfrokontr. Beret.*, 100, 88, 1971.
117. Kietreiber, M., Filter paper fluorescence test for detecting the presence of *Septoria nodorum* in *Triticum aestivum* taking into account seed in a dormant state, *Seed Sci. Technol.*, 9, 717, 1981.
118. Mathur, S. B. and Lee, S. L. N., A quick method for screening wheat seed samples for *Septoria nodorum*, *Seed Sci. Technol.*, 6, 925, 1978.
119. Safeeulla, K. M. and Shetty, H. S., Seed transmission of sorghum downy mildew in corn, *Seeds Farms*, 3(1), 21, 1977.
120. Jacobsen, B. J. and Williams, P. H., Histology and control of *Brassica oleracea* seed infection by *Phoma lingam*, *Plant Dis. Rep.*, 55, 934, 1971.
121. Klisiewicz, J. M., Wilt-incitant *Fusarium oxysporum* f. *carthami* present in seed from infected safflower, *Phytopathology*, 53, 1046, 1963.
122. Singh, T. and Sinclair, J. B., Histopathology of *Cercospora sojina* in soybean seeds, *Phytopathology*, 75, 185, 1985.
123. Kunwar, I. K., Singh, T., and Sinclair, J. B., Histopathology of mixed infections by *Colletotrichum truncatum* and *Phomopsis* spp. or *Cercospora sojina* in soybean seeds, *Phytopathology*, 75, 489, 1985.
124. Fialkovaskoya, E. A., *Loose Smut of Wheat*, Gossel' khozizdat ukr SSR, Kiev, 1963.
125. Bubentozoff, S., A method for the isolation of the smut fungus *Ustilago tritici* (Pers.) Rostr. from infected wheat grain, *Plant Prot. (Leningr.)*, 12, 89, 1937.
126. Wöstmann, E., The detection of *Ustilago tritici* in wheat grain by optical fluorescence, *Kuhn Arch.*, 6, 247, 1942.
127. Naumova, N. A., Application of ultraviolet rays for the detection of diseased seeds, *Plant Prot. (Moscow)*, 44, 1957.
128. Agarwal, V. K., Agarwal, M., Gupta, R. K., and Verma, H. S., Studies on loose smut of wheat. I. A simplified procedure for the detection of seed-borne infection, *Seed Res.*, 9, 49, 1981.
129. Morton, D. J., A quick method of preparing barley embryos for loose smut examination, *Phytopathology*, 50, 270, 1960.
130. Munjal, R. L. and Chatrath, M. S., Preliminary screening of wheat varieties against loose smut by the embryo test, *Indian Phytopathol.*, 27, 238, 1964.
131. Pedersen, P. N., A routine method of testing barley seed for loose smut (*Ustilago nuda* (Jens.) Rostr.), *Proc. Int. Seed Test. Assoc.*, 21, 1, 1956.
132. Popp, W., An improved method of detecting loose smut mycelium in whole embryo of wheat and barley, *Phytopathology*, 48, 641, 1958.
133. Russell, R. C., The whole embryo method of testing barley for loose smut as a routine test, *Sci. Agric.*, 30, 361, 1950.
134. Simmonds, P. M., Detection of the loose smut fungi in embryos of barley and wheat, *Sci. Agric.*, 26, 51, 1946.
135. Linder, D. H., An ideal mounting medium for mycologists, *Science*, 70, 430, 1929.
136. Skvortzoff, S. S., A simple method for detecting hyphae of loose smut in wheat grains, *Plant Prot. (Leningrad)*, 15, 90, 1937.
137. Boedijn, K. B., Trypan blue in a stain for fungi, *Stain Technol.*, 31, 115, 1956.
138. Laidlaw, M. R. M., Extracting barley embryos for loose smut examination, *Plant Pathol.*, 10, 63, 1961.

139. Rennie, W. J. and Seaton, R. D., Loose smut of barley — the embryo test as a means of assessing loose smut infection in seed stocks, *Seed Sci. Technol.*, 3, 697, 1975.

140. Hewett, P. D., A note on the extraction rate in the embryo method for loose smut of barley, *Ustilago nuda* (Jens.) Rostr., *Proc. Int. Seed Test. Assoc.*, 35, 181, 1970.

141. Agarwal, V. K., Verma, H. S., and Singh, S. B., Technique for the detection of loose smut infection in wheat seeds, *Seed Technol. News*, 8(3), 1, 1978.

142. Khanzada, A. K., Rennie, W. J., Mathur, S. B., and Neergaard, P., Evaluation of two routine embryo test procedures for assessing the incidence of loose smut infection in seed samples of wheat, *Seed Sci. Technol.*, 8, 363, 1980.

143. Kavanagh, T. and Mumford, D. L., Modification and adaptation of Popp's technique to routine detection of *Ustilago nuda* (Jens.) Rostr. in barley embryos, *Plant Dis. Rep.*, 44, 591, 1960.

144. Russell, R. C. and Popp, W., The embryo test as a method of forecasting loose smut infection in barley, *Sci. Agric.*, 31, 559, 1951.

145. Agarwal, V. K. and Srivastava, A. K., A simpler technique for routine examination of rice seed lots for rice bunt, *Seed Technol. News*, 11(3), 1, 1981.

146. Agarwal, V. K. and Verma, H. S., A simple technique for the detection of Karnal bunt infection in wheat seed samples, *Seed Res.*, 11, 100, 1982.

147. Guthrie, J. W., Routine methods for detecting and enumerating seed-borne bacterial plant pathogens, *J. Seed Technol.*, 4, 78, 1979.

148. Srivastava, D. N. and Rao, Y. P., Epidemic of bacterial blight of rice in North India, *Indian Phytopathol.*, 16, 390, 1963.

149. Cother, E. J., Bacterial glume blotch of rice, *Plant Dis. Rep.*, 58, 1126, 1974.

150. Burkholder, W. H., The bacterial blight of the bean: a systemic disease, *Phytopathology*, 11, 63, 1921.

151. Saettler, A. W. and Perry, S. K., Seed transmitted bacterial disease in Michigan navy (pea) beans *Phaseolus vulgaris*, *Plant Dis. Rep.*, 56, 378, 1972.

152. Burke, D. W., Incidence of bacterial pathogens in dry beans in irrigated districts of Nebraska, Wyoming and Colorado in 1954 and 1955, *Plant Dis. Rep.*, 41, 488, 1957.

153. Schuster, M. L. and Sayre, R. M., A coryneform bacterium induces purple-coloured seed and leaf hypertrophy of *Phaseolus vulgaris* and other Leguminosae, *Phytopathology*, 57, 1064, 1967.

154. Dunleavy, J., Stunt disease of soybean caused by *Corynebacterium* sp., *Phytopathology*, 52, 8, 1962.

155. Lai, M., Bacterial canker of bell pepper caused by *Corynebacterium michiganense*, *Plant Dis. Rep.*, 60, 339, 1976.

156. Parker, M. C. and Dean, L. L., Ultraviolet as a sampling aid for detection of bean seed infected with *Pseudomonas phaseolicola*, *Plant Dis. Rep.*, 52, 534, 1968.

157. Taylor, J. D., The quantitative estimation of the infection of bean seed with *Pseudomonas phaseolicola* (Burkh.) Dowson, *Ann. Appl. Biol.*, 66, 29, 1970.

158. Wharton, A. L., Detection of infection by *Pseudomonas phaseolicola* (Burkh.) Dowson in white-seeded dwarf bean seed stocks, *Ann. Appl. Biol.*, 60, 305, 1967.

159. Chakravarti, B. P. and Rangarajan, M., A virulent strain of *Xanthomonas oryzae* isolated from rice seeds in India, *Phytopathology*, 57, 688, 1967.

160. Strider, D. L., Detection of *Xanthomonas nigromaculans* f. sp. *zinniae* in zinnia seed, *Plant Dis. Rep.*, 63, 869, 1979.

161. Ednie, A. B. and Needham, S. M., Laboratory test for internally-borne *Xanthomonas phaseoli* var. *fuscans* in fieldbean (*Phaseolus vulgaris* L.) seed, *Proc. Assoc. Off. Seed Anal.*, 63, 76, 1973.

162. Taylor, J. D., Bacteriophage and serological methods for the identification of *Pseudomonas phaseolicola* (Burk.) Dowson, *Ann. Appl. Biol.*, 66, 387, 1970.

163. Schaad, N. W., Use of direct and indirect immunofluorescence tests for identification of *Xanthomonas campestris*, *Phytopathology*, 68, 249, 1978.

164. Schaad, N. W. and Kendrick, R., A qualitative method for detecting *Xanthomonas campestris* in crucifer seed, *Phytopathology*, 65, 1034, 1975.

165. Kado, C. I. and Heskett, M. G., Selective media for isolation of *Agrobacterium*, *Corynebacterium*, *Erwinia*, *Pseudomonas* and *Xanthomonas*, *Phytopathology*, 60, 969, 1970.

166. Hsich, S. P. Y., Buddenhagen, I. W., and Kauffman, H. E., An improved method for detecting the presence of *Xanthomonas oryzae* in rice seed, *Phytopathology*, 64, 273, 1974.

167. Wakimoto, S., Studies on the multiplication of OP₁ phage (*Xanthomonas oryzae* bacteriophage), *Sci. Bull. Fac. Agric. Kyushu Univ.*, 15, 151, 1955.

168. Gross, D. C. and Vidaver, A. K., A selective medium for isolation of *Corynebacterium nebraskense* from soil and plant parts, *Phytopathology*, 69, 82, 1979.

169. Schaad, N. W. and White, W. C., A selective medium for soil isolation and enumeration of *Xanthomonas campestris*, *Phytopathology*, 64, 876, 1974.

170. Lundsgaard, T., A new method for detection of *Xanthomonas campestris* in Brassica seeds, in *Proc. 15th Int. Workshop on Seed Pathology*, Paris, 1975, 14.

171. Saettler, A. W., Seedling injection as an aid in identifying bean blight bacteria, *Plant Dis. Rep.,* 55, 703, 1971.
172. Saettler, A. W., Detection of internally borne blight bacteria in navy (pea) bean seed, Proc. 2nd Int. Congr. Plant Pathology, St. Paul, Minn., 1973, 816.
173. Trujillo, G. E. and Saettler, A. W., A combined semi-selective medium and serology test for the detection of Xanthomonas blight bacteria in bean seed, *J. Seed Technol.,* 4, 35, 1979.
174. Parashar, R. D. and Leben, C., Detection of *Pseudomonas glycinea* in soybean seed lots, *Phytopathology,* 62, 1075, 1972.
175. Volcani, Z., A quantitative method for assessing cucumber seed infection caused by *Pseudomonas lachrymans, Isr. J. Bot.,* 15, 192, 1966.
176. Shackleton, D. A., A method for the detection of *Xanthomonas campestris* (Pammel 1885) Dowson 1939, in Brassica seed, *Nature (London),* 193, 78, 1962.
177. Srinivasan, M. C., Neergaard, P., and Mathur, S. B., A technique for detection of *Xanthomonas campestris* in routine seed health testing of crucifers, *Seed Sci. Technol.,* 1, 853, 1973.
178. Hunter, R. E. and Brinkerhoff, L. A., Longevity of *Xanthomonas malvacearum* on and in cotton seed, *Phytopathology,* 54, 617, 1964.
179. Singh, R. A. and Rao, M. H. S., A simple technique for detecting *Xanthomonas oryzae* in rice seeds, *Seed Sci. Technol.,* 5, 123, 1977.
180. Shekhawat, P. S. and Chakravarti, B. P., Comparison of agar plate and cotyledon methods for the detection of *Xanthomonas vesicatoria* in chilli seeds, *Phytopathol. Z.,* 94, 80, 1979.
181. Kennedy, B. W., Detection and distribution of *Pseudomonas glycinea* in soybean, *Phytopathology,* 59, 1618, 1969.
182. Humaydan, H. S., A bioassay for seed-borne *Xanthomonas campestris,* in Proc. 15th Int. Workshop on Seed Pathology, Paris, 1975, 15.
183. Thyr, B. D., Assaying tomato seed for *Corynebacterium michiganense, Plant Dis. Rep.,* 53, 858, 1969.
184. Venette, J., Disease and bean seed certification, *N.D. Seed J.,* 3, 1978.
185. Guthrie, J. W., Huber, D. M., and Fenwick, H. S., Serological detection of halo blight, *Plant Dis. Rep.,* 49, 297, 1965.
186. Guthrie, J. W., Testing bean seed for bacterial pathogens in Idaho, USA, *Seed Pathol. News,* 5, 3, 1973.
187. Elango, F. and Lozano, J. C., Sexual seed transmission of *Xanthomonas manihotis, Fitopatol. Colomb.,* 8, 15, 1979.
188. Ercolani, G. L., Efficiency and measurement of transmission of *Xanthomonas vesicatoria* and *Corynebacterium michiganense* in tomato, *Ind. Conserve,* 43, 15, 1968.
189. Sutton, M. D., Bacteriophages of *Pseudomonas atrofaciens* in cereal seeds, *Phytopathology,* 56, 727, 1966.
190. Sutton, M. D. and Katznelson, H., Isolation of bacteriophages for the detection and identification of some seedborne pathogenic bacteria, *Can. J. Bot.,* 31, 201, 1953.
191. Wakimoto, S., The determination of the presence of *Xanthomonas oryzae* by the phage technique, *Sci. Bull. Fac. Agric. Kyushu Univ.,* 14, 495, 1954.
192. Katznelson, H., The detection of internally borne bacterial pathogens of beans by a rapid phage plaque count technique, *Science,* 112, 645, 1950.
193. Katznelson, H. and Sutton, M. D., A rapid phage plaque method for the detection of bacteria as applied to the demonstration of internal bacterial infection of seed, *J. Bacteriol.,* 61, 689, 1951.
194. Randhawa, P. S. and Singh, N., Detection of *Xanthomonas campestris* pv. *malvacearum* from cotton seed by a modified phage multiplication method, in *Proc. 3rd Int. Symp. Plant Pathology,* New Delhi, 1981, 104.
195. Wallen, V. R., Sutton, M. D., and Grainger, P. N., A high incidence of fuscous blight in Sanilac beans from South Western Ontario, *Plant Dis. Rep.,* 47, 652, 1963.
196. Carroll, T. W., Methods of detecting seed-borne plant viruses, *J. Seed Technol.,* 3, 82, 1979.
197. Phatak, H. C., Seed-borne plant viruses — identification and diagnosis in seed health testing, *Seed Sci. Technol.,* 2, 3, 1974.
198. Phatak, H. C., Seed transmitted plant viruses — methods of testing, in *Seed Pathology — Problems and Progress,* Yorinori, J. T., Sinclair, J. B., Mehta, Y. R., and Mohan, S. K., Eds., Fundação Instituto Agronômico do Paraná, IAPAR, Londrina, Brazil, 1979, 147.
199. Koshimizu, V. and Iizuka, N., Studies on soybean virus diseases in Japan, *Bull. Tohoku Agric. Exp. Stn.,* 27, 1, 1963.
200. Kennedy, B. W. and Cooper, R. L., Association of virus infection with mottling of soybean seed coats, *Phytopathology,* 57, 35, 1967.
201. Ross, J. P., Interaction of the soybean mosaic and bean pod mottle viruses infecting soybeans, *Phytopathology,* 53, 887, 1963.

202. Almeida, A. M. R. and Miranda, L. C., Occurrence of soybean common mosaic virus in Parana State and its seed transmissibility, *Fitopatol. Bras.*, 4, 293, 1979.

203. Goodman, R. M., Bowers, G. R., Jr., and Paschal, E. H., II, Identification of soybean germplasm lines and cultivars with low incidence of soybean mosaic virus transmission through seed, *Crop Sci.*, 19, 264, 1979.

204. Ross, J. P., Effect of temperature on mottling of soybean seed caused by soybean mosaic virus, *Phytopathology*, 60, 1798, 1970.

205. Hill, J. H., Importance and detection of soybean mosaic virus in seed, *Iowa Seed Sci.*, 3, 8, 1981.

206. Owen, F. V., Hereditary and environmental factors that produce mottling in soybean, *J. Agric. Res.*, 34, 559, 1927.

207. Porto, M. D. M. and Hagedorn, D. J., Seed transmission of a Brazilian isolate of soybean mosaic virus, *Phytopathology*, 65, 713, 1975.

208. Wilcox, J. R. and Laviolette, F. A., Seed-coat mottling response of soybean genotypes to infection with soybean mosaic virus, *Phytopathology*, 58, 1446, 1968.

209. Hill, J. H., Lucas, B. S., Benner, H. I., Tachibana, H., Hammond, R. B., and Pedigo, L. P., Factors associated with the epidemiology of soybean mosaic virus in Iowa, *Phytopathology*, 70, 536, 1980.

210. Kuhn, C. W., Symptomatology, host range, and effect on yield of seed-transmitted peanut virus, *Phytopathology*, 55, 880, 1965.

211. Teploukhova, T. N. and Karimov, T. M., Necrosis of tomato seeds and its connection with tobacco mosaic virus, *Byull. Vses. Nauchno Issled. Inst. Zashch. Rast.*, 41, 60, 1977.

212. Middleton, J. T., Seed transmission of squash mosaic virus, *Phytopathology*, 34, 405, 1944.

213. Phatak, H. C. and Summanwar, A. S., Detection of plant viruses in seeds and seed stocks, *Proc. Int. Seed Test. Assoc.*, 32, 625, 1967.

214. Grogan, R. G. and Bardin, R., Some aspects concerning seed transmission of lettuce mosaic virus, *Phytopathology*, 40, 965, 1950.

215. Allam, E. K., Olfat, E. H., and Nagwa, A. M., Effect of virus infection on seed production and virus seed transmission of legumes. II. Cowpea aphid borne mosaic virus and its effect on cowpea seed, in Proc. 19th Int. Seed Testing Assoc. Congr., Vienna, 1980.

216. Allam, E. K., Olfat, E. H., and Nagwa, A. M., Effect of virus infection on seed production and virus seed transmissin of legumes. I. Bean common mosaic virus and its effect on bean seed, 19th Int. Seed Testing Assoc. Congr., Vienna, 1980.

217. Agarwal, V. K., Nene, Y. L., and Beniwal, S. P. S., Detection of bean common mosaic virus in urdbean (*Phaseolus mungo*) seeds, *Seed Sci. Technol.*, 5, 619, 1977.

218. Afanasiev, M. M., Occurrence of barley stripe mosaic in Montana, *Plant Dis. Rep.*, 40, 142, 1956.

219. Hampton, R. E., Sill, W. H., Jr., and Hansing, E. D., Barley stripe mosaic virus in Kansas and its control by a greenhouse seedlot testing technic, *Plant Dis. Rep.*, 41, 735, 1957.

220. McKinney, H. H., Culture methods of detecting seed-borne virus in Glacier barley seedlings, *Plant Dis. Rep.*, 38, 152, 1954.

221. Rohloff, I., Lettuce mosaic, *Proc. Int. Seed Test. Assoc.*, 30, 1065, 1965.

222. Rohloff, I., The controlled environment room test of lettuce seed for identification of lettuce mosaic virus, *Proc. Int. Seed Test. Assoc.*, 32, 59, 1967.

223. Rohloff, I., Lettuce mosaic, in *Handbook on Seed Health Testing Work Sheet No. 9*, 2nd ed., International Seed Testing Association, Zurich, 1980, 4.

224. Crowley, N. C., Studies on the seed transmission of plant virus diseases, *Aust. J. Biol. Sci.*, 10, 449, 1957.

225. Pelet, F., Dosage du virus de la mosaique de la laitue par indexage de la grainesur *Chenopodium quinoa*, *Rev. Hortic. Suisse*, 38, 7, 1965.

226. Pelet, F. and Gagnebin, F., Production de graines de laitue sans virus, *Rev. Hortic. Suisse*, 36, 22, 1963.

227. Taylor, R. H., Grogan, R. G., and Kimble, K. A., Transmission of tobacco mosaic virus in tomato seed, *Phytopathology*, 51, 837, 1961.

228. Athow, K. L. and Bancroft, J. B., Development and transmission of tobacco ringspot virus in soybean, *Phytopathology*, 49, 697, 1959.

229. Quantz, L., A plate test for the rapid demonstration of common bean mosaic virus (*Phaseolus* virus), *Nachrichtenbl. Dtsch. Pflanzenschutzdienstes (Braunschweig)*, 9, 71, 1957.

230. Quantz, L., Demonstrating bean common mosaic virus in bean seeds by means of dish test, *Nachrichtenbl. Dtsch. Pflanzenschutzdienstes (Braunschweig)*, 14, 49, 1962.

231. Hampton, R. O., Factors affecting detection of pea seedborne mosaic virus in *Pisum* seedlots, *Proc. Am. Phytopathol. Soc.*, 3, 207, 1976.

232. Chandra, K. J. and Summanwar, A. S., Quick methods for detection of cowpea mosaic virus in cowpea seeds, in Proc. 2nd Int. Symp. Plant Pathology, New Delhi, 1977, 52.

233. Walkey, D. G. A. and Whittingham-Jones, S. G., Seed transmission of strawberry latent ringspot virus in celery (*Apium graveolens* var. *Dulce*), *Plant Dis. Rep.*, 54, 802, 1970.

234. George, J. A., A technique for detecting virus infected Montmorency cherry seeds, *Can. J. Plant Sci.*, 42, 198, 1962.
235. Frosheiser, F. I., Virus infected seeds in alfalfa seed lots, *Plant Dis. Rep.*, 54, 591, 1970.
236. Carroll, T. W., Gossel, P. L., and Batchelor, D. L., Use of sodium dodecyl sulfate in serodiagnosis of barley stripe mosaic virus in embryos and leaves, *Phytopathology*, 69, 12, 1979.
237. Ball, E. M., *Serological Tests for the Identification of Plant Viruses,* American Phytopathological Society, St. Paul, Minn., 1974, 31.
238. Slack, S. A. and Shepherd, R. J., Serological detection of seed-borne barley stripe mosaic virus by a simplified radial diffusion technique, *Phytopathology*, 65, 948, 1975.
239. Hamilton, R. I., An embryo test for detecting seed-borne barley stripe mosaic virus in barley, *Phytopathology*, 55, 798, 1965.
240. Langenberg, W. G. and Ball, E. M., High pH ammonia agar immuno-diffusion for plant viruses, *Phytopathology*, 62, 1214, 1972.
241. Scott, H. A., Serological detection of barley stripe mosaic virus in single seeds and dehydrated leaf tissues, *Phytopathology*, 51, 200, 1961.
242. Carroll, T. W., Barley stripe mosaic virus: its importance and control in Montana, *Plant Dis.*, 64, 136, 1980.
243. Lima, J. A. A. and Purcifull, D. E., Immunochemical and microscopical techniques for detecting blackeye cowpea mosaic and soybean mosaic viruses in hypocotyls of germinated seeds, *Phytopathology*, 70, 142, 1980.
244. Hamilton, R. I. and Nichols, C., Serological methods for detection of pea seed-borne mosaic virus in leaves and weeds of *Pisum sativum, Phytopathology*, 68, 539, 1978.
245. Zimmer, R. C., Influence of agar on immunodiffusion serology of pea seedborne mosaic virus, *Plant Dis. Rep.*, 63, 278, 1979.
246. Cockbain, A. J., Bowen, R., and Vorra-Urai, S., Seed transmission of broadbean stain virus and echtes Ackerbohnenmosaik virus in fieldbean *(Vicia faba), Ann. Appl. Biol.*, 84, 321, 1976.
247. Lister, R. M., Application of the enzyme-linked immunosorbent assay for detecting viruses in soybean seed and plants, *Phytopathology*, 68, 1393, 1978.
248. Moreira, R. de A. and Perrone, J. C., Purification and partial characterization of a lectin from *Phaseolus vulgaris, Plant Physiol.*, 59, 783, 1977.
249. Jermoljev, E. and Chod, J., Pouzitiserologicke methody K testovani obesne mozaiky Fazulu na kliccich semen Fazolu, *Ochr. Rost.*, 2, 145, 1966.
250. Singer, J. M. and Plotz, C. M., The latex fixation test. I. Application to the serologic diagnosis of rheumatoid arthritis, *Am. J. Med.*, 21, 888, 1956.
251. Marcussen, O. F. and Lundsgaard, T., A new micromethod for the latex agglutination test, *Z. Pflanzenkr. Pflanzenschutz*, 82, 547, 1975.
252. Lundsgaard, T., Routine seed health testing for barley stripe mosaic virus in barley seeds using the latex-test, *Z. Pflanzenkr. Pflanzenschutz*, 83, 278, 1976.
253. Powell, C. C. and Schlegel, D. E., Factors influencing seed transmission of squash mosaic virus, *Phytopathology*, 60, 1466, 1970.
254. Clark, M. F. and Adams, A. N., Characteristics of the microplate method of enzyme linked immunosorbent assay for the detection of plant viruses, *J. Genet. Virol.*, 34, 475, 1977.
255. Voller, A. A., Bartlett, D. D., Bidwell, D. E., Clark, M. F., and Adams, A. N., The detection of viruses by enzyme-linked immunosorbent assay (ELISA), *J. Genet. Virol.*, 33, 165, 1976.
256. Bossennec, J. M. and Maury, Y., Problem and preliminary data concerning routine detection of soybean mosaic virus in soybean seeds, *Ann. Phytopathol.*, 9, 223, 1977.
257. Brlansky, R. H. and Derrick, K. S., Detection of seed-borne plant viruses using serologically specific electron microscopy, *Proc. Am. Phytopathol. Soc.*, 3, 334, 1976.
258. Jafarpour, B., Shepherd, R. J., and Grogan, R. G., Serologic detection of bean common mosaic and lettuce mosaic viruses in seed, *Phytopathology*, 69, 1125, 1979.
259. Casper, R., Virus assay of seed samples by ELISA, 3rd Int. Congr. Plant Pathology, P. Parey, Berlin, 1978, 111.
260. Derrick, K. S. and Brlansky, R. H., Assay for viruses and mycoplasmas using serologically specific electron microscopy, *Phytopathology*, 66, 815, 1976.
261. Brlansky, R. H. and Derrick, K. S., Detection of seedborne plant viruses, using serologically specific electron microscopy, *Phytopathology*, 69, 96, 1979.
262. Gold, A. H., Suneson, C. A., Houston, B. R., and Oswald, J. W., Electron microscopy and seed and pollen transmission of rod-shaped particles associated with the false-stripe virus disease of barley, *Phytopathology*, 44, 115, 1954.
263. Gardner, W. S., Electron microscopy of barley stripe mosaic virus: comparative cytology of tissues infected during different stages of maturity, *Phytopathology*, 57, 1315, 1967.

264. Shalla, T. A., Electron microscopy of cells infected with barley stripe mosaic virus as a result of mechanical and seed transmission, in *Viruses of Plants,* Beemster, A. B. R. and Dijkstra, J., Eds., North-Holland, Amsterdam, 1966, 94.

265. Walkey, D. G. A. and Webb, M. J. W., Tubular inclusion bodies in plants infected with viruses of the NEPO type, *J. Genet. Virol.,* 27, 159, 1970.

266. Schippers, B., Transmission of bean common mosaic virus by seed of *Phaseolus vulgaris* L. cultivar Beka, *Acta Bot. Neerl.,* 12, 433, 1963.

267. Couch, H. B. and Gold, A. H., Rod-shaped particles associated with lettuce mosaic, *Phytopathology,* 44, 715, 1954.

268. Mai, W. F. and Lyon, H. H., *Pictorial Key to Genera of Plant Parasitic Nematodes,* 4th ed., Cornell University Press, Ithaca, N. Y., 1975.

269. Ou, S. H., *Rice Diseases,* Commonwealth Mycological Institute, Kew, England, 1972, 368.

270. Nandakumar, C., Prasad, J. S., Rao, Y. S., and Rao, J., Investigations on the white tip nematode *(Aphelenchoides besseyi), Indian J. Nematol.,* 5, 62, 1975.

271. Goodey, T., *Anguillulina dipsaci* in the inflorescence of onions and samples of onion seed, *J. Helminthol.,* 21, 22, 1943.

272. Huu-Hai Vuong, The occurrence in Madagascar of the rice nematodes, *Aphelenchoides besseyi* and *Ditylenchus angustus,* in *Nematodes of Tropical Crops,* Tech. Commun. No. 40, Commonwealth Bureau of Helminthology, Kew, England, 1969, 274.

273. Thorne, G., *Principles of Nematology,* McGraw-Hill, New York, 1961, 553.

274. Dropkin, V. H., *Introduction to Plant Nematology,* John Wiley & Sons, New York, 1980, 293.

275. Tamura, I. and Kegasawa, K., Studies on the ecology of the rice nematode *Aphelenchoides besseyi* Christie. I. On the swimming away of rice nematode from the seeds soaked in water and its relation to the water temperature, *Jpn. J. Ecol.,* 7, 111, 1957.

276. Fenwick, D. W., A refinement of Gemmell's single cyst technique, *J. Helminthol.,* 21, 37, 1943.

277. Goodey, T., *Anguillulina dipsaci* on onion seed and its control by fumigation with methyl bromide, *J. Helminthol.,* 17, 21, 1945.

278. Ayoub, S. M., *Plant Nematology, an Agricultural Training Aid,* State of California Department of Food and Agriculture, Sacramento, 1977.

279. Southey, J. F., Laboratory methods for work with plant and soil nematodes, Tech. Bull. No. 2, Ministry of Agriculture, Fisheries and Food (Great Britain), Her Majesty's Stationery Office, London, 1970, 148.

Chapter 13

DETERIORATION OF GRAINS BY STORAGE FUNGI

Seeds are used for propagation of the species, as food for humans, or feed for animals, and are the result of a series of activities that include planting, harvesting, processing or conditioning, and storage. Deterioration of seeds can occur at any time from planting to final use either through the natural aging process or due to damage caused by microorganisms, viruses, insects, birds, or rodents. The theories concerned with the biochemical and physiological processes of seed deterioration were reviewed by Anderson.[1] The role of seed enzymes in seed deterioration was reviewed by Cherry[2] and St. Angelo and Ory.[3] Infection or infestation by microorganism and viruses can render seeds unfit for any of the basic uses. The involvement of bacteria, fungi, and viruses in seed deterioration was reviewed by Harman,[4] and that of fungi in seed storage by Christensen and Sauer.[5] Mills[6] reviewed the interactions between insects and fungi in seed deterioration. Deterioration resistance mechanisms were reviewed by Halloin.[7]

I. FIELD AND STORAGE FUNGI

Seed deterioration may be initiated by fungi, bacteria, or yeasts. The role of bacteria and yeasts in grain deterioration has not been investigated extensively.

Fungi that attack seeds are classified into either field or storage fungi on the basis of their ecological requirements.[5,8] Field fungi invade seeds either during development or after maturity, but before harvest. Generally, damage caused by field fungi occurs in the field with little or no damage occurring during storage. For field fungi to cause damage to occur during storage, moisture content must be in equilibrium with relative humidities above 95%, which gives a seed moisture content of 24 to 25% on a wet weight basis in seeds of barley, maize, oat, soybean, and wheat. This moisture content is far above that at which these seeds are stored; hence, field fungi rarely cause damage during storage.

Storage fungi are those which grow in grains and seeds when moisture content is in equilibrium with 70 to 80% relative humidities. Storage fungi normally do not play a role in disease development in the field but play a major role in seed deterioration in storage. They have the ability to grow without free water. Storage fungi mainly are species of *Aspergillus* and *Penicillium*.[5] *A. amstelodami, A. candidus, A. chevalieri, A. flavus, A. repens, A. restrictus, A. ruber,* and *A. ochraceus;* and *Penicillium* spp. usually invade grains with a moisture content above 16% and at low temperatures.[8] *P. viridicatum* is a common storage fungus which invades dent corn kernels stored at 19 to 24% moisture content between 8 to 24°C.[9] *Penicillium* spp. are more prevalent in temperate than in tropical regions. Monographs on *Aspergillus*[10] and *Penicillium*[11] deal with species identification. In addition, species of *Mucor* and *Rhizopus*,[12] *Absidia*,[13] *Byssochlamys*,[14] and *Candida* and *Hansenula*[15] have been reported on stored grains.

II. INVASION BY STORAGE FUNGI

Storage fungi generally deteriorate seeds after harvest when relative humidity and temperatures are high (Figure 46).[5] However, *Aspergillus flavus* colonizes maize silk tissue (stigma and style) and invades developing maize kernels. Insect feeding on maize kernels is not necessary for establishment of the fungus.[16] The interaction between cultivar and planting date that promotes silking during periods of high airborne spore loads of *A. flavus* promotes kernel infection.[17]

In peanut, germtubes from conidia of *A. glaucus* enter stalks through stomata and

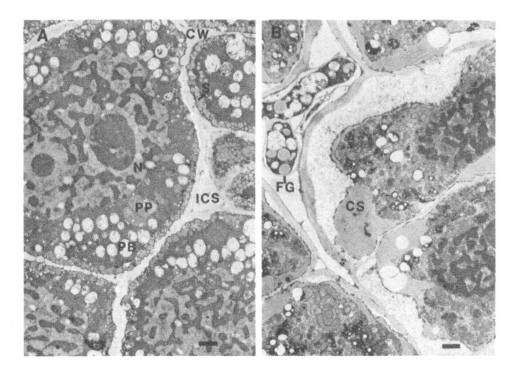

FIGURE 46.   Electron micrographs of wheat (*Triticum aestivum*) embryo tissue of (A) control and (B) storage fungi-infected seed showing cell wall (CW), nucleus (N), intercellular space (ICS), proplastids (PP), protein bodies (PB), spherosomes (S), coalesced spherosomes (CS), and fungal hyphae (FG). Bars = 1 μm. (From Anderson, J. D. and Baker, J. E., *Phytopathology*, 73, 321, 1983. With permission.)

pods through crevices caused by mechanical damage followed by colonization of the inner fruit wall. *A. halophilicus* invades seeds of Indian red pepper directly by production of appressoria and infection pegs.[18]

Conidia and other resting structures of storage fungi are present on equipment used in the seed industry, such as bags, combines, gums, and threshers, and in seed processing. Seeds are invaded during harvesting, threshing, post-threshing operations, and storage. The percentage of wheat seeds yielding *Aspergillus* spp. increases between harvest and arrival at a grain terminal. Threshed seeds are invaded by *Aspergillus* spp. more rapidly than seeds in ears. Inoculum is rare in the air around ripe wheat fields, moderately abundant in air around country elevators, where freshly harvested grain is being handled, and highly abundant in the air around a grain elevator.[19]

Inoculum of storage fungi may be present as dormant spores on the outside of grains, under the pericarp, or within seed tissues. The hyphae of storage fungi can be found within tissues of and on the inner side of barley hulls,[20] under the wheat pericarps,[21,22] and on the surface of wheat embryos.[23] The hyphae of several storage fungi are confined to pericarp layers in maize seeds, showing no visible signs or effects of invasion.[24]

Survival of storage fungi depends upon pathogen species, storage conditions, and many other factors. Members of the *A. glaucus* group persist for at least 11 months in soybean seeds stored at 6.5% moisture content, whereas *A. niger* does not.[25,26]

III. LOSSES

The role of storage fungi in grain deterioration has been underestimated. Approxi-

Table 13
LOSS OF SEED GERMINABILITY CAUSED BY STORAGE FUNGI

Crop	Moisture content (% wet wt)	Temp. (°C)	Storage period	Treatment	Germination (%)	Ref.
Pisum sativum	Not given	30	6 months	Noninoculated	97	29
				Inoculated with storage fungi	0	
Triticum aestivum	17.3—17.8	22—25	60—68 days	Noninoculated	100	30
				Inoculated with *Aspergillus ochraceus*	2	
	16—16.4	25	2 months	Noninoculated	90	31
				Inoculated with *A. amstelodami, A. candidus,* and *A. restrictus*	27	
Zea mays	17—18	15	2 years	Noninoculated	96	32
				Inoculated with storage fungi	0	
	19.1—19.9	27—32	74 days	Noninoculated	97	33
				Inoculated with *A. flavus*	13	

mately 4% of all stored grain is lost annually due to storage fungi.[27] Losses can be especially high under tropical conditions where high humidity, rainfall, and temperature coupled with poor storage conditions exist. It was not until 1960, after 100,000 turkeys died due to the intake of *Aspergillus flavus*-invaded peanuts, that the role of saprophytic fungi, such as *Aspergillus* and *Penicillium* spp., in grain deterioration became known. The major losses due to storage fungi are as follows.

A. Decrease in Germinability

A high, uniform germination is desired if seeds are to be used for planting or malting, or as edible sprouts. Decreased germinability can be caused by mechanical or physiological reasons or by storage fungi. If caused by storage fungi, the effect on germination is influenced by the moisture content of the seeds, period of storage, storage temperature, and other factors.

Storage fungi which invade seeds are classified generally as saprophytes, but experimental evidence has shown that most storage fungi invade seed embryos preferentially and reduce germination. Seed samples of barley, maize, peas, sorghum, soybean, and wheat stored at moisture contents and temperatures favorable for growth of storage fungi, but free of fungi, retain a germinability of 95 to 100% for some months, whereas with similar samples inoculated with storage fungi, germinability is reduced to zero or near zero (Table 13).[28]

Rice seed samples stored at 25 and 31°C with a 14% moisture content decrease in germinability and increase in storage fungi (*A. candidus, A. glaucus,* and *A. restrictus*) proportional to increasing storage time.[34] Sunflower seeds stored at moisture contents of 10, 12, and 14% and at 3 to 5, 8 to 10, and 27 to 28°C and infested with *Alternaria* spp., *Aspergillus glaucus,* and *Penicillium* spp. decreased in germinability proportional to increasing moisture content, temperature, and storage time.[35] Maize seeds invaded with *A. flavus* at 19 to 20% moisture content and 20 to 25°C had 13% germination whereas noninoculated seeds had 97% germination after 74 days.[33] Maize seeds free of storage fungi stored at 17% moisture content and 25°C retained a 98% germination after 12 weeks whereas those inoculated with storage fungi germinated at 6%.[36] *A. candidus,* species of the *A. glaucus* group, and *Penicillium* spp. reduced barley seed

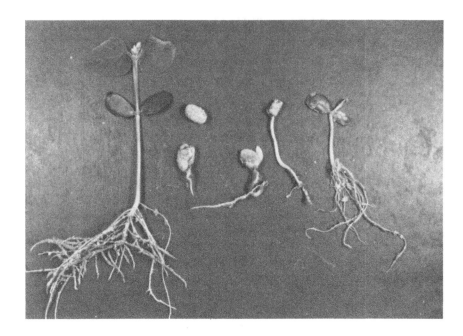

FIGURE 47.   Reduced germination and vigor of soybean (soyabean) (*Glycine max*) seedlings from seeds infected with *Aspergillus melleus* (right), compared to a noninfected seedling (left). (From Sinclair, J. B., Ed., *Compendium of Soybean Diseases,* American Phytopathological Society, St. Paul, Minn., 1982, 75. With permission.)

germination.[20,37] Sorghum seeds with a moisture content of 14.5% on a wet weight basis, and invaded by storage fungi, had a decreased germinability proportional to the increased moisture content and storage period.[38] A higher recovery of *A. flavus* and *A. melleus* was correlated with decreased germination in soybean (Figure 47).[39,40] *P. oxalicum* reduced germination in maize seeds infected at the full-silk stage.[41]

Reduced germination can vary depending upon the species of *Aspergillus* involved. Pea seeds stored at 85% relative humidity and 30°C were killed in 3 months when inoculated with *A. flavus,* in 6 months with *A. candidus* and *A. ruber,* and in 8 months with *A. restrictus,* whereas noninoculated seeds maintained a 97% germination for 6 months.[29]

Loss in germination due to storage fungi may be attributed to several factors. A toxin produced by *A. ruber* kills tissues in embryonic axes of pea seeds in advance of infection.[42] Wheat seeds infected with *Aspergillus* spp., when imbibed with water, become a jelly-like mass, suggesting that cell-wall degrading enzymes may be involved. In contrast, pea and squash embryos are killed without physical invasion by fungi, indicating the involvement of diffusable toxins.[43] Mitochondria isolated from the embryonic axes of *A. ruber*-infected pea seeds were less active than those from noninfected seeds suggesting that mitochondria damage by *A. ruber* plays a role in seed deterioration.[44]

B. Discoloration and Shrinkage of the Grains

Aspergillus and *Penicillium* spp. cause discoloration and shrinkage of barley seeds.[37] Maize kernels artificially inoculated with *A. candidus* and stored at 18% moisture content and at 25°C develop a dark brown discoloration at the germ ends after 4.5 months, whereas noninoculated seeds appear normal without loss in germination.[45] Wheat seeds inoculated with *A. candidus* with 16.0 to 16.4% moisture content and stored for 3 months at 25°C showed 79% seeds with dark germ ends and 6% germination compared to 95% germination of normal germ end noninoculated seeds.[31]

C. Heating

The temperature of stored grain always increases due to seed respiration. In addition, respiration by insects and storage fungi also can increase storage temperature. The rate of respiration, and thus the rise in temperature, are higher in seeds invaded by storage fungi. This has been shown for soybeans[46] and wheat.[47,48] A temperature increase from 50 to 55°C and a parallel increase in respiration was associated with proliferation of *A. flavus* and *A. glaucus* in soybeans.[49] Temperature increases accompanied by increased water content in barley seeds from 15 to 24% favored development of storage fungi.[37]

It is difficult to separate increases in temperature due to storage fungi and insects. In temperate zones or wet tropics, temperature increases are due to fungi which develop in moist seeds, but in the dry tropics, insect-induced dry-grain heating is possible.[50] Temperatures may rise to above 60°C when induced by fungi, whereas insect-induced hot spots never exceed 46°C. Hot zones generally rise vertically through the bulk when induced by fungi, and expand laterally when induced by insects because of the insect mobility.[50]

D. Spoilage in Nutritive Value

Storage fungi use stored grains as a substrate resulting in the chemical breakdown of nutrients.

1. Increase in Fatty Acid Value (FAV)

Grain deterioration is accompanied by an increase in fatty acid value (FAV), and the FAV level in any given sample provides an indication of the stage of deterioration and grain storability. When the FAV increases, the odor and flavor of fatty acids makes the grain rancid. The level of fatty acids produced varies depending upon species and strains within a species of storage fungi.

An increase in fatty acid was correlated with the presence of storage fungi in maize[51] and wheat.[52] The FAV of sound maize kernels is low but increases with increasing levels of damaged kernels[53] and invasion by fungi.[54,55] The increase in FAV may be due to action of seed lipases or microflora on the lipids in grains.[55] FAV increases more rapidly in broken maize kernels inoculated with *A. glaucus* and *A. restrictus* stored at 15.5% moisture content and 30°C than in broken kernels free of fungi and similarly stored.[56]

2. Biochemical Changes in Nutritional Value

There is a continuous loss in protein content of cowpea seeds infested by *A. flavus.*[57] This may be due to hydrolysis of seed protein by hydrolytic enzymes as was reported for peanuts infected by *A. parasiticus.*[58]

E. Production of Toxins

Storage fungi may produce mycotoxins which are injurious to man and animals on consumption.

IV. CONDITIONS FAVORING STORAGE FUNGI DEVELOPMENT

The invasion and successful establishment of, and seed deterioration by storage fungi is influenced by a number of factors which may act alone or in combination.

Table 14
MOISTURE CONTENT NECESSARY FOR GROWTH OF
STORAGE FUNGI[26]

	Crop		
			Triticum aestivum (wheat);
	Glycine max	*Sorghum vulgare*	*Zea mays*
Fungus	(soyabean, soybean)	(milo, sorghum)	(corn, maize)
Aspergillus restrictus	12.0—12.5	14.0—14.5	13.5—14.5
A. glaucus	12.5—13.0	14.0—15.0	14.0—14.5
A. candidus	14.5—15.0	16.0—16.5	15.0—15.5
A. ochraceus	14.5—15.0	16.0—16.5	15.0—15.5
A. flavus	17.0—17.5	19.0—19.5	18.0—18.5
Penicillium sp.	16.0—18.5	17.0—19.5	16.5—19.0

Table 15
MOISTURE CONTENT, WET WEIGHT BASIS, OF COMMON GRAINS
AND SEEDS AT EQUILIBRIUM WITH RELATIVE HUMIDITIES OF 65
TO 85% AT 20 TO 25°C[5,28]

		Crop			
		Helianthus annuus	*Oryza sativa* (paddy, rice)		*Triticum aestivum* (wheat);
Relative humidity (%)	*Glycine max* (soyabean, soybean) (%)	(sunflower) anchenes[a] (%)	Rough (%)	Polished (%)	*Zea mays* (corn, maize) (%)
65	12.5	8.0	12.5	14.0	12.5—13.5
70	13.0	9.0	13.5	15.0	13.5—14.0
75	14.0	10.0	14.0	15.5	14.5—15.0
80	16.0	11.0	15.0	16.5	16.0—16.5
85	18.0	13.0	16.5	17.5	18.0—18.5

Note: The figures are approximate; the equilibrium moisture content of a given kind of seed at a given relative humidity will vary with several factors.

[a]   Oil type; the confectionary types have equilibrium moisture contents about 1% higher.

A. Moisture Content

Moisture content of grain prior to storage is the most important factor for establishment, development, and growth of storage fungi during storage. Various storage fungi have different moisture content requirements below which they fail to develop or may remain dormant with seeds without causing any damage (Table 14).[25,26] The role of moisture content in grain invasion has been reviewed.[5,8]

Generally, storage fungi grow at moisture contents in equilibrium with relative humidities of 65 to 90%. The moisture content of some common seeds in equilibrium with relative humidities within this range are presented in Table 15.[5,28] Each species within a given genus of a storage fungus has a lower limit of relative humidity below which it will not grow (Table 16).[28] The upper relative humidity limit is determined in part by the nature of the fungus and in part by competition.[28] The quantity of physically bound "free water" within cereals generally determines the storability of seeds.

A moisture content lower than 13% retards development of all types of microorganisms in seeds, and below 10% insects fail to develop. The moisture content within

Table 16
MINIMUM RELATIVE HUMIDITY
THAT PERMITS GROWTH OF
VARIOUS STORAGE FUNGI

Fungus	Min. relative humidity (%)
Aspergillus halophilicus	65
A. restrictus	70
A. repens	73
A. candidus, A. ochraceus	80
A. flavus	85
Penicillium spp.	85—95

From Christensen, C. M., *Seed Sci. Technol.*, 1, 547, 1973. With permission.

a seed lot may be different in different pockets and also may change from season to season.[59] The lower limit of moisture content that permits invasion of sunflower seeds by *Aspergillus glaucus* and *A. restrictus* is about 6%, and at moisture contents below 6.5%, invasion is very low.[60] *A. flavus* does not invade maize seeds stored with moisture contents below 17.5% wet weight basis but does invade at 18.5% and above. At 18.5% moisture content, 25 and 35°C favor invasion by *A. ochraceus* and *A. candidus*, respectively.[33]

An increase in moisture content from 15 to 24% and temperature from 20 to 30°C in barley results in an increase of *Aspergillus* and *Penicillium* spp. and decreased germination.[37]

B. Temperature

Storability and development of storage fungi within grains are influenced by atmospheric, grain, and intergranular air temperature. Atmospheric heat slowly enters the grain bulk. In addition, heat also is produced by fungi, insects, and other organisms which is considerably higher than the heat produced by seeds. Most storage fungi do not develop below 0°C, mites below 5°C and insects below 15°C. The minimum, optimum, and maximum temperature requirement for the growth of most storage fungi is 0 to 5, 30 to 33, and 50 to 55°C, respectively. Low temperature can be used as a substitute for control of high moisture content because at temperatures below 10°C storage fungi that invade seeds at moisture contents in equilibrium with relative humidities up to 85% will grow very slowly.[28] Some *Penicillium* spp. grow slowly at −5°C when the moisture content is in equilibrium with a relative humidity of 90% or more.[28] Papavizas and Christensen[61] found that wheat seed with a moisture content of up to 10% can be stored without deterioration for 1 year at 10°C or below; and with a moisture content of up to 18% for 19 months at −5°C. Rice seeds with an initial moisture content of 12 to 14% were stored without deterioration for 465 days at 5 and 15°C.[34] Soybean seeds stored at 15°C germinated above 95% after 24 weeks at 12.1 to 16.5% moisture content.[62]

C. Physical Damage of the Seed

Damaged or injured seeds are more susceptible than whole seeds to invasion by storage fungi and to deterioration by these molds after invasion.[63] Care in threshing procedures to eliminate or reduce seed damage helps preserve germinability in wheat seeds stored under moist conditions. Germinability of machine-threshed seeds decreases faster and results in fewer living seeds at 80, 85, and 90% relative humidity than hand-threshed seeds.[63] A high level of injury on barley seeds reduces germination faster than a low level of injury.[37] Loss of germinability in damaged maize seeds is greater than in nongerminated seeds.[32] Loss of germination in damaged wheat seeds inoculated with *Penicillium* spores is greater than in undamaged seeds.[64]

D. Degree of Seed Infestation/Invasion Prior to Storage

Seeds infected in the field are likely to deteriorate faster because storage fungi continue to develop at lower moisture and temperature conditions.[28] Maize kernels invaded by storage fungi prior to storage deteriorate more rapidly under conditions favorable to the fungi than kernels free of storage fungi.[32]

E. Admixtures with the Seed

Any admixture with seeds such as plant parts, broken seeds, weed seeds, soil, or field insects such as grasshoppers and crickets, can serve as a substrate for growth of storage fungi or insects. Admixtures usually are more moisture absorbent and retentive, and thus more susceptible to storage-fungi colonization. Weed seeds may have a higher moisture content than cultivated crop seeds.[28] Sound, clean seeds are deteriorated less by storage fungi than uncleaned seeds.

F. Length of Storage

Length of storage influences storage fungi. The chances of damage are less for grains stored for a few weeks or months compared to grains stored for years at a given moisture or temperature safe for storage. Moisture content and temperature are likely to change due to metabolic activities of storage fungi and insects; hence, it becomes essential to periodically check the condition of stored grain.

G. Insect and Mite Infestation

Infestation of grain by insects and mites can accelerate deterioration by storage fungi. Insects and mites can carry fungal spores thus infesting previously noninfested grains. In addition, they increase the moisture content of the grain through the release of water from their digestive processes.[5,28,65] Respiration releases carbon dioxide and water.[5,28,65] Cereal grains infested by the granary weevil, *Sitophilus granarius*, and stored for 3 months at 75% relative humidity showed an increase in moisture content from 12.1% to between 17.6 and 23.0% compared to 14.6 to 14.8% in weevil-free grains.[66] The increase in storage fungi was correlated with insect infestation.[67,68] The mites, *Acarus siro* and *Tyrophagus castellanii*, carry *Aspergillus glaucus* spores on their bodies and in their digestive tract and feces.[69] An increase in moisture content by the gram moth, *Sitotroga cerealella*, is responsible for increased populations of *A. amstelodami*, *A. repens*, and *A. ruber*.[68] Bottles of wheat infested with *Aeroglyphus robustus* mites had higher levels of *Aspergillus glaucus* infection than did those containing *Lepidoglyphus destructor*.[70]

V. DETECTION OF DAMAGE

Grain should be checked at regular intervals during storage for damage due to storage fungi in order to take preventive measures. Damage to grains may be detected by several methods.

A. Grain Condition

Grain damage by invasion of storage fungi show one or more of the following symptoms: weakening of the embryo, embryo death, discoloration of embryo or seed, mustiness, and complete decay.[28] Discoloration of the germ may be detected by removing the pericarp. Discoloration may be complete or restricted to the tip of the grain or seed. Such germs are likely to be moldy and later turn dark brown.

B. Isolation of Fungi

Isolation of storage fungi will provide information about the organism(s) involved in grain deterioration:

Agar media — Seeds are plated on an agar medium for isolation of fungi. Christensen and Qasem[71] used a medium containing 18% common salt to distinguish *Aspergillus restrictus* from *A. glaucus*. T-6 (25 g Difco tomato juice agar, 15 g Difco powdered agar, 60 g sodium chloride, 900 m*l* distilled water) is suitable for detection of storage fungi. This medium restricts fungal growth so that fungi from adjacent seeds do not overgrow one another.[28] Coconut agar is useful for quantitative identification of *A. flavus* isolates that produce aflatoxin.[72] Surface-sterilized grains should be used for most fungal isolations.

Blotter method — Moist blotters are used for detection of storage fungi. *Penicillium* sp. grows well on barley seeds after incubation on moist filter paper without surface disinfection.[73] Filter paper soaked with water and 7.5% common salt was found superior for detection of *A. glaucus* and *Penicillium* spp.[74]

Bright-field microscopic examination — Microscopic examination of barley grains was used to show an increase in mycelial growth with increasing moisture content (15 to 24%) and rising temperature (20 to 30°C).[37]

C. Observation Under UV Light

A characteristic bright greenish-yellow fluorescence under UV (≥365 nm) is seen on dead maize seeds when *A. flavus* is paired with other fungi.[75]

D. Measurement of Gases

The measurement of CO_2 production is used to determine losses in seed quality and to predict storability of maize.[76] The production of 7.4 g CO_2 per kilogram of dry matter of maize kernels results in about 0.5% loss in dry matter.[77] At moisture contents which permit fungal growth, three major gases are involved — butanol, ethanol, and methanol — in maize and soybean seeds. Their production is correlated with fungal growth as measured by chitin content of seeds and plating on agar media.[78]

E. Determination of FAV

Increases in the FAV are due to the activities of storage fungi. Normally, a grain with a low FAV is preferred over one with a high FAV. However, deterioration may occur without an increase in FAV.

F. Moldy Smell

A heavy mycelial growth in barley usually is accompanied by a moldy odor, but such an odor may occur without visible mycelial growth.[37]

G. Collection of Seed Exudates

Exudates from germinating seeds are influenced by temperature, soil moisture, injuries, dormancy, and oxygen tension. The simplest way of collecting the exudates from seeds is to place 10 to 12 g of surface-disinfested seeds in 15 to 25 m*l* sterile distilled water in a sterile container. The amount of water is just sufficient for seed imbibition and germination initiation. After the desired interval, remove the liquid and wash the seeds in a small amount of distilled water and pool the wash water with liquid collected earlier. The maximum collection period should not exceed the time it takes

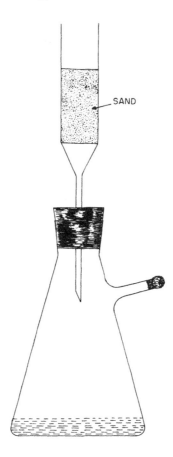

FIGURE 48.   Schematic diagram of the system for collecting of
exudates from a single seed. (After Hayman, D. S., *Can. J. Bot.,*
47, 1521, 1969. From Dhingra, O. D. and Sinclair, J. B., *Basic
Plant Pathology Methods,* CRC Press, Boca Raton, Fla., 1985.
With permission.)

for the radicle to reach the length of the seed. At the time of collection check the
sterility of the solution by streaking a loopful on nutrient agar.[79-82]

Loria and Lacy[83] placed individual seeds in moist sand in 25-mm diameter test tubes
(cotton plugged and autoclaved) and covered them with sterile sand. After germination, the seeds were removed aseptically, rinsed in the same tube with sterile distilled
water, and placed on potato-dextrose agar (PDA) to check the sterility. The sand from
each tube is washed with distilled water to obtain the exudate solution.

Hayman[80,81] developed a simple system that permits repeated collection of exudates
from a single seed at the desired time intervals without disturbing or damaging the
germinating seeds. A glass tube (5 cm long and 2.2 cm diameter), to one end of which
a 3.5-cm piece of capillary tubing is sealed, is half filled with sand (Figure 48). The
free end of the capillary is fitted into a rubber stopper in a 25-m*l* Erlenmeyer flask
with side arm and sealed with silicon. The sand tube and the side arm are plugged with
cotton. Several assemblies thus are prepared and autoclaved. A single surface-disinfested seed is placed at uniform depth in each tube. At predetermined intervals, the
exudates are collected by rinsing the sand with 10 m*l* of sterile distilled water in two
portions of 5 m*l* each. The exudates are collected into the flask below by applying a
mild vacuum.

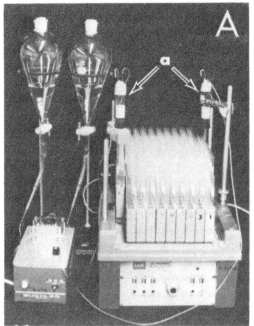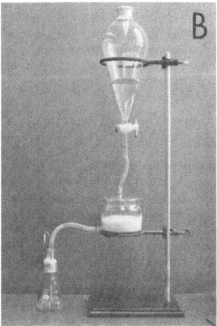

FIGURE 49. Sterile leaching systems for collection of exudates from a single seed during germination. (A) Periplastic pump delivers a predetermined amount of water per hour to the substratum in which the seed is germinating, with the leachings being collected every hour; (B) exudates from a number of seeds germinating in a substratum in a modified culture are collected every 8 hr. (From Short, G. E. and Lacy, M. L., *Phytopathology*, 66, 182, 1976; and Dhingra, O. D. and Sinclair, J. B., *Basic Plant Pathology Methods*, CRC Press, Boca Raton, Fla., 1985. With permission.)

The collection of exudates by continuous leaching has been used by Schlub and Schmitthenner.[84] A buchner funnel (40 mm diameter) is filled with 5 mℓ of 1-mm glass beads supported on a fine mesh nylon screen. The mouth of the funnel is covered with an aluminum cap containing four holes and a glass tube sealed with tygon tubing to deliver 3 mℓ of water per hour for leaching. The assembly is autoclaved and a number of seeds are buried halfway into the beads and then covered with 9 mℓ of 3-mm glass beads. The leachates are collected in test tubes containing a 1-mℓ solution of 500 μg/mℓ each of streptomycin sulfate and chlorophenicol.

Short and Lacy[85] collected exudates from seeds for 1 hr (Figure 49). A glass tube, 90 mm long and 25 mm diameter with a rubber stopper on each end is filled with 20 g of 1-mm glass beads. A separatory funnel is connected to the tube with a tygon tubing. The bottom rubber stopper is fitted with a 7-mm diameter drain hole. Following sterilization of the apparatus, sterile distilled water is poured aseptically into the funnel and a surface-disinfested seed is placed within the glass bead matrix. The glass cylinder may be covered with aluminum foil to exclude light. The water is then percolated through the glass bead matrix at 10 mℓ/hr. A fraction collector is used to collect the leachates at 1-hr intervals in test tubes containing 10 mℓ of a 50% ethanol to control microbial growth.

VI. CONTROL

The losses due to deterioration of grains by storage fungi can be reduced by the following methods.

A. Avoiding Damage to Seeds during Harvesting, Processing, and Threshing

Storage fungi gain entrance to seeds at any stage after maturity to harvest and during processing and threshing. Any damage to seeds results in the entry of storage fungi and poses a problem during storage. Such seeds are likely to carry a heavy invasion of fungi. Proper precautions at harvest and during postharvesting operations help reduce occurrence of storage fungi. Seeds which are clean, sound, undamaged, and dried are ideal for storage.

B. Storage Conditions

The most effective means of avoiding damage to grains by storage fungi is to maintain storage conditions that will prevent fungi development.[28] Grain samples should be examined periodically for moisture content, germination, and fungi. Any damage, if detected early, may be checked by taking suitable precautions. This may be achieved by drying seeds to a safer moisture level limit before storage and then frequent aeration during storage to maintain safe moisture and temperature limits below which the storage fungi do not develop. Aeration is the most important method to uniformly reduce grain temperature to 5 to 10°C throughout the storage bin. At such temperatures, storage fungi grow slowly and insects and mites are dormant. Aeration also lowers moisture content of grains, but grains, being hygroscopic, gain moisture produced by insects, mites, and fungi and absorb atmospheric moisture. Periodic aeration lowers the temperature and removes moisture to a limit safe for storage. Storage bins with aeration keep the incidence of storage fungi at a low level thus avoiding seed damage. Navarro et al.[86] found that wheat seeds maintain 90% germination after 22 months storage in a metal bin equipped with an aeration system. Grain drying can be achieved using high-speed dryers, solar energy, unheated air, or partially heated air. The latter methods result in physically better grain but there are more chances of infection by storage fungi.

C. Reducing Seed Moisture to Safe Limits

The most effective means of avoiding grain damage is to dry grains to a moisture content at which no storage fungi will develop. Damage by storage fungi will not occur in storage bins at moisture contents below 13% for starchy cereal seeds, such as barley, maize, rice, sorghum, and wheat; below 12% for soybeans; and below 10% for flax seeds.[5] Low moisture content is a key factor in maintaining seed viability in storage. The ability of the container to minimize moisture gain during storage is as important as low initial moisture content in preserving seed viability in the tropics. Unprotected seeds absorb moisture from the air until equilibrium is reached. This is especially critical in the tropics where wide, daily fluctuations of relative humidity occur. The key to preserving seed viability under the humid conditions in the tropics is to dry the seed to a level of moisture content below which no storage fungi will develop. For soybeans, to avoid deterioration, seeds are dried to 8.6% moisture content on a dry weight basis and stored in a container which minimizes moisture gain from the atmosphere while in storage.[59] To avoid damage to sunflower seeds, the moisture content should be less than 6.5%.[60] For barley, a moisture content of 15% at 20°C retains germination up to 41 months.[37] Rapid moisture equilibration between wet and dry maize indicates that seeds can be blended with minimal risk of mold damage or aflatoxin contamination if the average moisture content of the blend is low enough to prevent mold growth.[87]

D. Seed Treatment

A number of chemicals reduce development of storage fungi. An ideal chemical must have low mammalian toxicity and long-lasting microbial-inhibiting properties.[76] If the grain is to be used as planting seed, the chemical should not have an adverse effect on

Table 17

APPLICATION RATES OF
PROPIONIC ACID IN ACCORDANCE
WITH DIFFERENT MOISTURE
CONTENTS

Moisture content (%)	Application rate (%)	Crop: cereal, *Zea mays* (corn, maize) (g/100 kg)
20	0.5	500
25	0.7—0.8	700—800
30	1.1—1.2	1100—1200
35	1.4—1.5	1400—1500
40	1.7—1.8	1700—1800
45	1.0—2.1	2000-2100

Note: The figures apply to undiluted acid and fresh cereal or maize.

From Sauer, D. B., Hodges, T. O., Burroughs, R., and Converse, H. H., *Trans. Am. Soc. Agric. Eng.*, 18, 1162, 1975. With permission.

germination. Matz and Milner[88] found that an addition of 0.2% or more of propylene oxide to wheat seeds checks deterioration of damp wheat but results in a heavy reduction in germination. Herting and Drury[89] reported that isobutyric acid was effective in controlling the fungi but treated grains failed to germinate.

Propionic acid salts are being used by bakers to prevent mold development. It prevents development of bacteria, molds, and yeast. Also, it blocks the activity of enzymes which decompose carbohydrates. It has no adverse toxicological effects. It also checks the formation of aflatoxin and overheating of wet cereals. In maize at 30% moisture content, propionic acid at 9.5 g/kg eliminates *Aspergillus* and *Penicillium* spp. up to 120 days.[90] The rate of infection by *A. candidus* was reduced significantly with sodium propionate (5000 μg/mℓ) in rice seeds (12.7 to 13% moisture) stored for 6 months at 75% and for 4 months at 85% relative humidity. It suppressed *A. candidus* but not *A. glaucus*.[91]

Propionic acid (1000 μg/mℓ) is used to control fungi in undried stored cereals and pulses destined for animal feed in temperate countries.[92] Acetic and formic acids are less effective but these are used in combination with propionic acid.[93] Methylene bispropionate is equal or superior to propionic acid. It breaks down into formaldehyde and propionic acid.[94,95] There are fungi and bacteria that have varying degrees of tolerance to propionic acid and other organic acids; therefore, a combination of two or more chemicals may be useful.[88] The rate of application of propionic acid depends upon the moisture content and storage period (Table 17).[96]

Chemicals which must be dissolved in water may not be useful against storage fungi since these fungi invade and cause damage at moisture contents in equilibrium with relative humidities of 70 to 90%.[97]

REFERENCES

1. Anderson, J. D., Deterioration of seeds during aging, *Phytopathology*, 73, 321, 1983.
2. Cherry, J. P., Protein degradation during seed deterioration, *Phytopathology*, 73, 317, 1983.
3. St. Angelo, A. J. and Ory, R. L., Lipid degradation during seed deterioration, *Phytopathology*, 73, 315, 1983.
4. Harman, G. E., Mechanisms of seed infection and pathogenesis, *Phytopathology*, 73, 326, 1983.
5. Christensen, C. M. and Sauer, D. B., Microflora, in *Storage of Cereal Grains and Their Products*, Christensen, C. M., Ed., American Association of Cereal Chemists, St. Paul, Minn., 1982, 219.
6. Mills, J. T., Insect-fungus associations influencing seed deterioration, *Phytopathology*, 73, 330, 1983.
7. Halloin, J. M., Deterioration resistance mechanisms in seeds, *Phytopathology*, 73, 335, 1983.
8. Christensen, C. M. and Kaufmann, H. H., Deterioration of stored grains by fungi, *Annu. Rev. Phytopathol.*, 3, 69, 1965.
9. Caldwell, R. W., William, W. C., and John, T., The occurrence and toxicity of Indian isolates of *Penicillium viridicatum*, in 2nd Int. Congr. Plant Pathol., St. Paul, Minn., 1973, 410.
10. Raper, K. B. and Fennell, D. I., *The Genus Aspergillus*, Williams & Wilkins, Baltimore, 1965, 686.
11. Pitt, J. I., *The Genus Penicillum*, Academic Press, New York, 1979, 634.
12. Clarke, J. H., Fungi in stored products, *PANS*, 15, 473, 1969.
13. Ellis, J. J. and Hesseltine, C. W., Species of *Absidia* with ovoid sporangiospores. II., *Sabourandia*, 5, 59, 1966.
14. Brown, A. H. S. and Smith, G., The genus *Paecilomyces* Bainier and its perfect stage *Byssochlamys* Westling, *Trans. Br. Mycol. Soc.*, 40, 17, 1957.
15. Wickerham, L. J., Taxonomy of Yeasts, Tech. Bull. No. 1029, U.S. Department of Agriculture, Washington, D.C., 1951, 56.
16. Jones, R. K., Duncan, H. E., Payne, G. A., and Leonard, K. J., Factors influencing infection by *Aspergillus flavus* in silk inoculated corn, *Plant Dis.*, 64, 859, 1980.
17. Jones, R. K., Duncan, H. E., and Hamilton, P. B., Planting date, harvest date, and irrigation effects on infection and aflatoxin production by *Aspergillus flavus* in field corn, *Phytopathology*, 71, 810, 1981.
18. Seenappa, M., Stobbs, L. W., and Kempton, A. G., *Aspergillus* colonization of Indian red pepper during storage, *Phytopathology*, 70, 218, 1980.
19. Tuite, J. F. and Christensen, C. M., Grain storage studies. XXIII. Time of invasion of wheat seed by various species of *Aspergillus* responsible for deterioration of stored grain, and source of inoculum of these fungi, *Phytopathology*, 47, 265, 1957.
20. Tuite, J. F. and Christensen, C. M., Grain storage studies. XVI. Influence of storage conditions upon the fungus flora of barley seed, *Cereal Chem.*, 32, 1, 1955.
21. Hyde, M. B., The subepidermal fungi of cereal grains. I. A survey of the world distribution of fungal mycelium in wheat, *Ann. Appl. Biol.*, 37, 179, 1950.
22. Christensen, C. M., Fungi on and in wheat seed, *Cereal Chem.*, 28, 408, 1951.
23. Christensen, C. M., Grain storage studies. XVIII. Mold invasion of wheat stored for sixteen months at moisture contents below 15 per cent, *Cereal Chem.*, 32, 107, 1955.
24. Russell, G. H., Murray, M. E., and Berjak, P., Storage microflora: on the nature of the host/pathogen relationship in fungal infected maize seeds, *Seed Sci. Technol.*, 10, 605, 1982.
25. Kennedy, B. W., Moisture content, mold invasion, and seed viability in stored soybeans, *Phytopathology*, 54, 771, 1964.
26. Kennedy, B. W., The occurrence of *Aspergillus* spp. on stored seeds, in *Seed Pathology — Problems and Progress*, Yorinori, J. T., Sinclair, J. B., Mehta, Y. R., and Mohan, S. K., Eds., Fundação Instituto Agronômico do Paraná, IAPAR, Londrina, Brazil, 1979, 257.
27. Anon., PL480 research stresses pest control and marketing, *Foreign Agric.*, 6, 6, 1968.
28. Christensen, C. M., Loss of viability in storage: microflora, *Seed Sci. Technol.*, 1, 547, 1973.
29. Fields, R. W. and King, T. H., Influence of storage fungi on the deterioration of stored pea seed, *Phytopathology*, 52, 336, 1962.
30. Christensen, C. M., Invasion of stored wheat by *Aspergillus ochraceus*, *Cereal Chem.*, 39, 100, 1962.
31. Papavizas, G. C. and Christensen, C. M., Grain storage studies. XXIX. Effect of invasion by individual species of *Aspergillus* upon germination and development of discoloured germs in wheat, *Cereal Chem.*, 37, 197, 1960.
32. Qasem, S. A. and Christensen, C.M., Influence of various factors on the deterioration of stored corn by fungi, *Phytopathology*, 50, 703, 1960.
33. Lopez, L. C. and Christensen, C. M., Effect of moisture content and temperature on invasion of stored corn by *Aspergillus flavus*, *Phytopathology*, 57, 588, 1967.
34. Fanse, H. A. and Christensen, C. M., Invasion by storage fungi of rough rice in commercial storage and in the laboratory, *Phytopathology*, 60, 228, 1970.

35. Christensen, C. M., Factors affecting invasion of sunflower seeds by storage fungi, *Phytopathology,* 59, 1699, 1969.
36. Moreno, E., Lopez, L. C., and Christensen, C. M., Loss of germination of stored corn from invasion by storage fungi, *Phytopathology,* 55, 125, 1965.
37. Welling, B., Fungus flora and germination of barley, *Tidsskr. Planteavl.,* 73, 291, 1969.
38. Christensen, C. M., Moisture content, moisture transfer and invasion of stored sorghum seeds by fungi, *Phytopathology,* 60, 280, 1970.
39. Dhingra, O. D., Nicholson, J. F., and Sinclair, J. B., Influence of temperature on recovery of *Aspergillus flavus* from soybean seed, *Plant Dis. Rep.,* 57, 185, 1973.
40. Ellis, M. A., Ilyas, M. B., and Sinclair, J. B., Effect of cultivar and growing region on internally seedborne fungi and *Aspergillus melleus* pathogenicity in soybean, *Plant Dis. Rep.,* 58, 332, 1974.
41. Caldwell, R. W., Tuite, J., and Carlton, W. W., Pathogenicity of Penicillia to ear rots, *Phytopathology,* 71, 175, 1981.
42. Harman, G. E. and Nash, G., Deterioration of stored pea seed by *Aspergillus ruber:* evidence for involvement of a toxin, *Phytopathology,* 62, 209, 1972.
43. Harman, G. E. and Pfleger, F. L., Pathogenicity and infection sites of *Aspergillus* species in stored seeds, *Phytopathology,* 64, 1339, 1974.
44. Harman, G. E. and Drury, R. E., Respiration of pea seeds (*Pisum sativum*) infected with *Aspergillus ruber, Phytopathology,* 63, 1040, 1973.
45. Qasem, S. A. and Christensen, C. M., Influence of moisture content, temperature, and time on the deterioration of stored corn by fungi, *Phytopathology,* 48, 544, 1958.
46. Milner, M. and Geddes, W. F., Grain storage studies. II. The effect of aeration, temperature, and time on the respiration of soybeans containing excessive moisture, *Cereal Chem.,* 22, 484, 1945.
47. Carter, E. P., Role of fungi in the heating of moist wheat, Circ. No. 838, U.S. Department of Agriculture, Washington, D.C., 1950.
48. Hummel, B. C. W., Cuendet, L. S., Christensen, C. M., and Geddes, W. F., Grain storage studies. XIII. Comparative changes in respiration, viability, and chemical composition of mold-free and mold-contaminated wheat upon storage, *Cereal Chem.,* 31, 143, 1954.
49. Milner, M. and Geddes, W. F., Grain storage studies. III. The relation between moisture content, mold growth, and respiration of soybeans, *Cereal Chem.,* 23, 225, 1946.
50. Howe, R. W., Loss of viability of seed in storage attributable to infestations of insects and mites, *Seed Sci. Technol.,* 1, 563, 1973.
51. Nagel, C. M. and Semeniuk, G., Some mold-induced changes in shelled corn, *Plant Physiol.,* 22, 20, 1947.
52. Milner, M., Christensen, C. M., and Geddes, W. F., Grain storage studies. VII. Influence of certain mold inhibitors on respiration of moist wheat, *Cereal Chem.,* 24, 507, 1947.
53. Baker, D., Neustadt, M. H., and Zeleny, L., Application of the fat acidity test as an index of grain deterioration, *Cereal Chem.,* 34, 226, 1957.
54. Bottomley, R. A., Christensen, C. M., and Geddes, W. F., Grain storage studies. X. The influence of aeration, time and moisture content on fat acidity, non-reducing sugars and mold flora of stored yellow corn, *Cereal Chem.,* 29, 53, 1952.
55. Christensen, C. M., Some changes in No. 2 corn stored two years at moisture contents of 14.5 and 15.2 percent and temperatures of 12, 20 and 25C, *Cereal Chem.,* 44, 95, 1967.
56. Sauer, D. B. and Christensen, C. M., Some factors affecting increase in fat acidity values in corn, *Phytopathology,* 59, 108, 1969.
57. Vijaya Kumari, P. and Karan, D., Deterioration of cowpea seeds in storage by *Aspergillus flavus, Indian Phytopathol.,* 34, 222, 1981.
58. Cherry, J. P., Young, C., and Beuchatt, L. A., Changes in proteins and total amino acids of peanuts (*Arachis hypogaea*) infested with *Aspergillus parasiticus, Can. J. Bot.,* 53, 2639, 1975.
59. Tenne, F. D., Ravalo, E. J., Sinclair, J. B., and Rodda, E. D., Changes in viability and microflora of soybean seeds stored under various conditions in Puerto Rico, *J. Agric. Univ. P.R.,* 62, 255, 1978.
60. Christensen, C. M., Moisture content of sunflower seeds in relation to invasion by storage fungi, *Plant Dis. Rep.,* 56, 173, 1972.
61. Papavizas, G. C. and Christensen, C. M., Grain storage studies. XXVI. Fungus invasion and deterioration of wheats stored at low temperatures and moisture contents of 15 to 18 percent, *Cereal Chem.,* 35, 27, 1958.
62. Dorworth, C. E. and Christensen, C. M., Influence of moisture content, temperature, and storage time upon changes in fungus flora, germinability and fat acidity values of soybeans, *Phytopathology,* 58, 1457, 1968.
63. Kulik, M. M., Retention of germinability and invasion by storage fungi of hand-threshed and machine-threshed wheat seeds in storage, *Seed Sci. Technol.,* 1, 805, 1973.
64. Hurd, A. M., Seedcoat injury and viability of seeds of wheat and barley as factors in susceptibility to molds and fungicides, *J. Agric. Res.,* 21, 99, 1921.

65. Christensen, J. J., Variability in the microflora in barley kernels, *Plant Dis. Rep.,* 47, 635, 1963.
66. Agarwal, N. S., Christensen, C. M., and Hodson, A. C., Grain storage fungi associated with the grain weevil, *J. Econ. Entomol.,* 50, 659, 1957.
67. Christensen, C. M. and Hodson, A. C., Development of granary weevils and storage fungi in columns of wheat. II., *J. Econ. Entomol.,* 53, 375, 1960.
68. Misra, C. P., Christensen, C. M., and Hodson, A. C., The angoimois grain moth, *Sitotroga cerealella* and storage fungi, *J. Econ. Entomol.,* 54, 1032, 1961.
69. Griffiths, D. A., Hodson, A. C., and Christensen, C. M., Grain storage fungi associated with mites, *J. Econ. Entomol.,* 52, 514, 1959.
70. White, N. D. G., Henderson, L. P., and Sinha, R. N., Effects of infestations by three stored-product mites on fat acidity, seed germination, and microflora of stored wheat, *J. Econ. Entomol.,* 72, 763, 1979.
71. Christensen, C. M. and Qasem, S. A., Detection of *Aspergillus restrictus* in stored grain, *Cereal Chem.,* 39, 68, 1962.
72. Holtmeyer, G. and Wallin, J. R., Identification of aflatoxin producing atmospheric isolates of *Aspergillus flavus, Phytopathology,* 70, 325, 1980.
73. Jørgensen, J., Changes in the microflora of barley seed stored with a high moisture content, *Tidsskr. Planteavl.,* 74, 425, 1970.
74. Mills, J. T., Sinha, R. N., and Wallace, H. A. H., Multivariate evaluation of isolation techniques for fungi associated with stored rape seed, *Phytopathology,* 68, 1520, 1978.
75. Wicklow, D. T. and Hesseltine, C. W., Fluorescence produced by *Aspergillus flavus* in association with other fungi in autoclaved corn kernels, *Phytopathology,* 69, 589, 1979.
76. Tuite, J. and Foster, G. H., Control of storage diseases of grain, *Annu. Rev. Phytopathol.,* 17, 343, 1979.
77. Steele, J. L., Saul, R. A., and Hukill, W. V., Deterioration of shelled corn as measured by carbon dioxide production, *Trans. ASAE,* 12, 685, 1969.
78. Donald, W. W. and Mirocha, C. J., Volatiles of grain storage fungi, *Proc. Am. Phytopathol. Soc.,* 1, 104, 1974.
79. Brookhouser, L. W. and Weinhold, A. R., Induction of polygalacturonase from *Rhizoctonia solani* by cotton seed and hypocotyl exudates, *Phytopathology,* 69, 599, 1979.
80. Hayman, D. S., The influence of temperature on the exudation of nutrients from cotton seeds and on preemergence damping-off by *Rhizoctonia solani, Can. J. Bot.,* 47, 1663, 1969.
81. Hayman, D. S., A note on the quantitative determination of carbohydrate in exudate from single cotton seeds and its significance, *Can. J. Bot.,* 47, 1521, 1969.
82. Kraft, J. M. and Erwin, D. C., Stimulation of *Pythium aphanidermatum* by exudates from mung bean seeds, *Phytopathology,* 57, 866, 1967.
83. Loria, R. and Lacy, M. L., Mechanism of increased susceptibility of bleached pea seeds to seed and seedling rot, *Phytopathology,* 69, 573, 1979.
84. Schlub, R. L. and Schmitthenner, A. F., Effects of soybean seed coat cracks on seed exudation and seedling quality in soil infested with *Pythium ultimum, Phytopathology,* 68, 1186, 1978.
85. Short, G. E. and Lacy, M. L., Carbohydrate exudation from pea seeds: effect of cultivar, seed age, seed color, and temperature, *Phytopathology,* 66, 182, 1976.
86. Navarro, S., Donahaye, E., and Calderson, M., Observations on prolonged grain storage with forced aeration in Israel, *J. Stored Prod. Res.,* 5, 73, 1969.
87. Sauer, D. B. and Burrough, R., Fungal growth, aflatoxin production and moisture equilibrium in mixtures of wet and dry corn, *Phytopathology,* 70, 516, 1980.
88. Matz, S. A. and Milner, M., Inhibition of respiration and preservation of damp wheat by means of organic chemicals, *Cereal Chem.,* 28, 196, 1951.
89. Herting, D. C. and Drury, E. E., Antifungal activity of volatile fatty acids on grains, *Cereal Chem.,* 51, 74, 1974.
90. Dhanraj, K. S., Chohan, J. S., Sunar, M. S., and Sona Lal, Preliminary studies in the use of luprosil in the prevention of microbial spoilage of wet maize during storage, *Indian Phytopathol.,* 26, 63, 1973.
91. Schroeder, H. W., Sodium propionate and infra-red drying for control of fungi infecting rough rice (*Oryza sativa*), *Phytopathology,* 54, 858, 1964.
92. Kozakiewicz, Z. and Clarke, J. H., Toxicity of propionic acid to some pre-harvest and post-harvest fungi of stored grain, in 2nd Int. Congr. Plant Pathology, St. Paul, Minn., 1973, 60.
93. Hall, G. E., Hill, L. D., Hatfield, E. E., and Jensen, A. H., Propionic-acetic acid for high moisture corn preservation, *Trans. Am. Soc. Agric. Eng.,* 17, 379, 1974.
94. Bothast, R. J., Black, L. T., Wilson, L. L., and Hatfield, E. E., Methylene-bis propionate preservation of high moisture corn, *J. Anim. Sci.,* 46, 484, 1978.
95. Sauer, D. B., Hodges, T. O., Burroughs, R., and Converse, H. H., Comparison of propionic acid and methylene bis propionate as grain preservatives, *Trans. Am. Soc. Agric. Eng.,* 18, 1162, 1975.

96. Zwick, W., A new conservation method for tapioca and maize, in Proc. Indian Science Congr. Kharagpur, 1972, 8.

Chapter 14

CONTROL OF SEEDBORNE PATHOGENS

I. INTRODUCTION

The principles of plant disease control have been defined and discussed in many texts available in personal or public libraries and will not be elaborated here. A flow chart summarizing the general classification of plant disease control methods is presented in Figure 50. Generally, no single method, except for immunity, will provide complete control for any one disease. The control of most plant diseases involves the use of more than one measure in a disease-control strategy similar to programs developed by entomologists in integrated pest-management systems.

Control of seedborne pathogens and diseases is attained through integrated disease-management systems. For example, the production of wheat seeds free of loose smut infection in India is attempted through an integrated approach using the following guidelines:[1] (1) production of seeds of relatively resistant cultivars; (2) production of seeds in areas isolated (about 150 m) from commercial plots; (3) field inspections to meet the requirements for certification, i.e., maximum permissible infection of 0.1% and 0.5% for foundation and certified seed production plots, respectively; (4) roguing infected plants; (5) testing of seeds using the embryo-count method to identify heavily infected seed lots; and (6) seed treatment with systemic fungicides.

The seedborne inoculum of *Leptosphaeria maculans* (*Plenodomus lingam* [*Phoma lingam*]) in *Brassica oleracea* was reduced to nondetectable levels[2] in the U.S. using the following: (1) seed production in the summer in dry areas of the West Coast; (2) use of resistant cultivars; (3) crop rotation; (4) reduction of residues; (5) seed health testing; and (6) seed treatment with benomyl.

Control of a seedborne pathogen is not likely to result in control of a disease if the pathogen survives by alternative means or if attempts at control are directed only toward control of seedborne inoculum.

In this chapter, methods which directly assist in prevention of seed infection in the field or control of seedborne infection and infestation are discussed.

II. SELECTION OF SEED PRODUCTION AREAS

Seeds should be produced in areas where the pathogens of major concern are unable to establish or maintain themselves at critical levels during periods of seed development. Areas with low rainfall and low relative humidity generally are favorable for production of high-quality seeds with low inoculum levels. Walker[3] identified western Washington State as an area for producing cabbage (*Brassica oleracea*) seeds. The Skagit Valley produces approximately 80% of the cabbage seed required for the U.S. and 30% for the world.[4] This moderately cool maritime climate, usually with dry summers, is ideal for production of high-quality cabbage seed relatively free of *Xanthomonas campestris* pv. *campestris*, and *Plenodomus lingam*.[5] However, two blackleg (*P. lingam*) epiphytotics in the eastern and mideastern growing areas occurred in 1947 and 1973 due to infected seeds produced in western Washington in 1946 and 1972, respectively. The epiphytotic in 1946 was due to the introduction of a virulent strain of *P. lingam* and wet weather at pod set, and in 1972 to the introduction of a new hybrid. An integrated disease-control approach to produce relatively pathogen-free seeds is now in effect and includes[5] seed health testing, benomyl seed treatment of infected stock seeds, and rotation of seed beds and production fields.

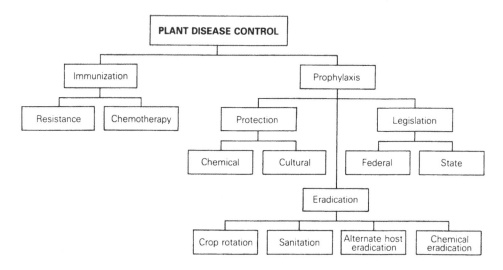

FIGURE 50.   General classification of various methods for the control of plant diseases. (After Sharvelle, E. G., *Plant Disease Control*, AVI Publishing, Westport, Conn., 1979, 311; and Tarr, S. A. J., *Principles of Plant Pathology*, Macmillan, London, 1972, 632.)

The Piedmont area is favored over the coastal plains of Georgia for seed production of cottonseed because low rainfall during boll opening results in reduced infection by *Alternaria alternata, Diplodia gossypina, Fusarium oxysporum, F. roseum, Glomerella gossypii,* and *Nigrospora sphaerica.*[6] Seed infection by *Septoria nodorum* is less when wheat is produced in the mountain region than in the southern part of Georgia.[7] *Ascochyta pinodella, A. pisi,* and *Mycosphaerella pinodes* occur frequently in pea (*Pisum sativum*) seed samples produced in the eastern U.S. (especially New England) and Canada, but seed stocks produced in the Palouse districts of the states of Idaho and Washington generally are free of these fungi and are commonly used throughout the U.S.[8] Because of low rainfall, pea seeds devoid of *Ascochyta* are produced from infected seeds when grown in the Imperial and Temecula Valleys of southern California.[8]

Sugar beet seed lots produced along the Pacific Coast of North America or in the Mediterranean region of Europe are unlikely to carry *Phoma betae.*[9] A high level of *Alternaria zinniae* occurs in *Zinnia elegans* seeds produced in coastal valleys of California, while those from the dry interior valleys may be free from the pathogen.[10] Bean seeds free of *Pseudomonas syringae* pv. *phaseolicola* are produced under semiarid conditions in the state of Idaho where there is zero tolerance of the bacterium in seed production fields.[11,12]

Lettuce seeds free of LMV are obtained annually from the Swan Hill area of Victoria and New South Wales, Australia since 1954. Vectors of the virus are not found during most of the growing season because of unfavorable climate and ecological conditions. The seeds produced in the Swan Hill area are used in the southern Victoria districts where the spread of mosaic from infected seeds has been a problem.[13] Peanut seeds free of PMV are produced for Georgia farmers by growing breeder seed under insect-free conditions in screenhouses where individual plants are indexed and rogued. Seeds then are grown in isolation from other peanuts or in a location free of the aphid vectors.[14]

III. CROP MANAGEMENT

Proper crop management can help in the production of relatively pathogen-free seeds.

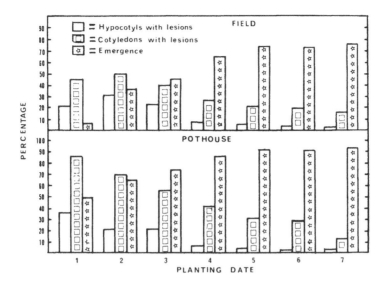

FIGURE 51.   Percentage emergence of soybean (soyabean) (*Glycine max*) seeds infected with *Colletotrichum truncatum* (anthracnose) planted on seven planting dates in pothouse and field tests during the monsoon (rainy) season in India, 1972, and percentage of hypocotyls and cotyledons of resulting plants that showed lesions. (From Nicholson, J. F. and Sinclair, J. B., *Plant Dis. Rep.*, 57, 770, 1973. With permission.)

A. High-Quality Seed

Planting seeds should be as free of the pathogen as possible. Seeds should come from a carefully maintained, genetically pure block and should be grown in selected production fields. They should be cleaned commercially and treated either chemically or nonchemically.

B. Seeding Rate

The seeding rate should not be excessive. A high seeding rate can increase the number of focuses of primary infection that may develop in the field which ultimately can result in higher disease incidence and seed infection.

C. Planting Time

Plant at a time when requirements of host and pathogen do not correspond and thus the plants may escape infection. Winter wheat sown early in autumn may escape infection from *Tilletia caries* and *T. foetida* because plants pass the susceptible stage before bunt spores germinate. Crops sown later may become infected.[15] Planting oat seeds in early spring reduces the incidence of *Ustilago avenae*. Germination of infected oat seeds at 7°C considerably reduces smut incidence, and at 0 to 2.5°C, reduction in smut is equivalent to using seed disinfection. In this temperature range the fungus does not develop.[16]

In India and Puerto Rico, for example, adjusting planting schedules so that soybeans mature at the end of (or out of) the rainy season reduces the amount of seedborne *Colletotrichum truncatum* (Figure 51). Although seed yields are lower than those produced during the rainy season, seed quality is superior.[17]

D. Burning

Burning over grass seed production fields in the state of Oregon destroys inoculum of *Gloeotinia temulenta* (blind seed disease) and *Claviceps purpurea* (ergot).[15]

E. Balanced Fertility

Adequate, balanced soil fertility, coupled with near neutral pH, is important in reducing seed infection. Plants under stress from deficient or toxic levels of nutrients are more susceptible to disease than those grown in soil with well-balanced fertility. Insufficient phosphorus or potash in soybean can increase losses from bacterial blight, bacterial pustule, charcoal rot, pod and stem blight, soybean cyst nematode, and several root and stem decay pathogens.[17] An excessive application of fertilizer may result in a greater disease incidence.

The incidence of seedborne *Alternaria padwickii* in rice increases proportionally to increased nitrogen from 0 to 200 kg/ha. Above 100 kg/ha, the incidence of *Curvularia lunata*, *Phoma sp.*, and *Trichothecium* sp. increased compared to 0 and 50 kg/ha.[18]

F. Planting Method

Planting method is effective in reducing seed transmission of TMV in tomato. A low transmission was obtained in direct planting (5%) as compared to transplanting (71%).[19] Sowing wheat seeds on the surface of recently flooded land results in a low incidence of flag smut (0.08 to 0.2%); in plots irrigated after sowing at 4-cm deep, a moderate incidence results (2.4 to 3.2%), and in soil moist enough for ploughing there is a high incidence (8.1 to 8.6%).[20] Rice seeded in water is less infested with the nematode *Aphelenchoides besseyi* than that drilled in and flooded when 6- to 9-cm high. Quiescent nematodes probably revive in water, move about, and die in the water before seed emergence.[21]

G. Spacing

Reduced spacing between plants favors seedborne infections. Close spacing results in high humidity among plants which can be conducive for heavy seedborne infections. At 15 cm between rice plants, there was a higher percentage of seedborne *Alternaria alternata*, *A. longissima*, *A. padwickii*, *Arthrobotrytis* sp., *C. lunata*, *Drechslera oryzae*, *Fusarium semitectum*, and *Sclerotium* sp. than at wider distances.[18] Soybean seeds from narrow-row (25 cm) compared to wide-row (76 cm) spacing give higher recovery of total fungi and bacteria which adversely affect the quality of clean seeds.[22]

H. Depth of Planting

Depth of planting greatly influences seed transmission of smuts. Shallow planting in wet soils protects wheat plants from *Urocystis tritici* (flag smut) in wheat.[20]

I. Water Management

Water management practices influence disease development.[17] Irrigation can be timed to reduce stress. Water management strategies vary depending upon growing area, most common diseases, soil types, etc. Irrigation, especially at the seed-development stage, may favor seed infection. Irrigation time and amount of water should be controlled so that the relative humidity is not raised to such an extent that it becomes conducive for seed infection. Overhead irrigation should be avoided where possible.

J. Crop Rotation

Crop rotation and clean tillage play an important role in controlling seedborne pathogens because many important bacterial and fungal pathogens survive between crops on or in crop debris. Rotating soybeans with a nonhost crop every second year is effective for reducing most foliage and stem pathogens, and rotation every third year sharply reduces soybean cyst nematodes.[17] Recommended tillage practices and rotation lengths vary with region and soil type. Soybean seed infection by *Phomopsis* sp. can be reduced by rotating soybean fields with maize.[15]

K. Isolation Distances

The distance between seed production and commercial plots has been worked out for reducing seedborne loose smut of barley and wheat. The distance between plots may vary from region to region depending on weather conditions. In Canada, the problem of proper isolation of stock seed production areas for barley is governed by legislation.[23] Similar legislation exists in Germany for stock seed production of barley and wheat; seed crops are rejected if loose smut is encountered in a field within a distance of less than 50 m of the seed field and if upwind of the prevailing wind direction.[24] In Holland, Oort[25] demonstrated that a minimum distance of 100 m from infected fields must be secured for barley grown under seed certification programs. Barley and wheat crops should be isolated by at least 50 m from any source of loose smut infection for production of certified seeds in the U.K. Seed Certification Scheme.

L. Roguing

Roguing should be practiced where possible. In seed production fields, infected plants should be rogued and destroyed. Roguing has been followed successfully in the control of loose smut of barley and wheat.

M. Foliar Fungicide Sprays

Fungicide sprays are used to control bacteria and fungi that cause disease on foliage and grains of field crops. Many of the microorganisms that cause disease of roots, stems, leaves, and reproductive structures are seedborne. The control of plant diseases in the field may indirectly or directly affect seed infection, but the information on the efficacy of fungicidal sprays on seed quality and infestation and/or infection by various mycoflora is limited to only few crops.

1. Soybean

The strongest evidence that fungicide sprays improve seed quality by controlling seedborne pathogens come from soybeans.[26] Prasartsee et al.[27] found that soybean seeds from plots sprayed with benomyl 50 WP + zinc + maneb 80 WP, thiophanate methyl 70 WP, benomyl 50 WP, chlorothalonil, zinc + maneb 80 WP, and thiabendazole 98.5 WP had significantly ($P = 0.05$) less seedborne *Phomopsis* than those from nonsprayed plants. Similarly, Ellis et al.[28-30] showed that soybean plants grown in Illinois and sprayed with benomyl produce seeds with significantly ($P = 0.05$) less *Phomopsis* seed decay than nonsprayed plants and that occurrence of *Phomopsis* was significantly correlated ($r = -0.71$) with germination in culture. Benomyl sprays reduced losses due to this disease even more so after delayed harvest.[31] In Brazil, fungicide sprays with benomyl, captafol, thiabendazole, chlorothanlonil, and cercobin significantly ($p = 0.05$) reduced seed coat and embryo infection by *Phomopsis* spp., and germination was signficantly higher for seeds from sprayed plants than nonsprayed plants.[32] Benomyl or a zinc ion-maneb complex sprayed on soybeans in Puerto Rico also controlled seedborne *Cercospora kikuchii*.[33] The incidence of most fungi in soybean seeds, *Diaporthe phaseolorum* var. *sojae*, *C. kikuchii*, *Colletotrichum truncatum*, and *Glomerella glycines* are reduced in seeds by two sprays of benomyl (1.1 kg/ha).[34]

The plant growth stage in which fungicides are applied is important in the control of seedborne infection (Figure 52; Table 18). Seed infection by *Phomopsis* spp. in soybean usually occurs after the R_7 growth stage, although pods could be infected much earlier.[35] The most effective time to apply fungicide is in the R_6 to R_7 growth stage, before seed infection takes place.[36] Benomyl applied at 1.2 g/l, either 3 days before or on the day of inoculation reduces *Phomopsis* pod and seed infection; when applied 3 days after inoculation, it is less effective.[36] Fungicide application just before seed infection takes place ensures an optimum concentration of methylbenzimidazole carba-

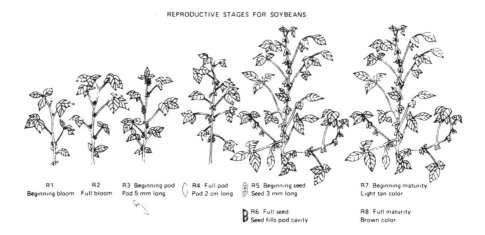

FIGURE 52.   Reproductive stages of soybean (soyabeans) (*Glycine max*). (After Walla, W. J., Ed., *Soybean Diseases Atlas*, Texas A & M University, College Station, Tex., 1979.) (From Sinclair, J. B., Ed., *Compendium of Soybean Diseases*, American Phytopathological Society, St. Paul, Minn., 1982, 2. With permission.)

Table 18
GROWTH STAGE KEY FOR SOYBEANS

Stage	Description[a]
V_1	Completely unrolled leaf at the unifoliolate node
V_2	Completely unrolled leaf at the first node above the unifoliolate node
V_3	Three nodes on main stem beginning with the unifoliolate node
V^N	N nodes on the main stem beginning with unifoliolate node
R_1	One flower at any node
R_2	Flower at node immediately below the uppermost node with a completely unrolled leaf
R_3	Pod 0.5-cm ($^1/_4$ in.) long at one of the four uppermost nodes with a completely unrolled leaf
R_4	Pod 2-cm ($^3/_4$ in) long at one of the four uppermost nodes with a completely unrolled leaf
R_5	Beans beginning to develop (can be felt when the pod is squeezed) at one of the four uppermost nodes with a completely unrolled leaf
R_6	Pod green containing full-size beans at one of the four uppermost nodes with a completely unrolled leaf
R_7	Pods yellowing, 50% of leaves yellow, physiological maturity
R_8	95% of pods brown, harvest maturity

[a]   Use the key to indicate the stage of soybean growth. Determine the vegetative stages by counting the number of nodes on the main stem, beginning with the unifoliolate node, which has a completely unrolled leaf.

After Fehr, W. R., Caviness, C. E., Burmood, D. T., and Pennington, J. S., *Crop Sci.*, 11, 929, 1971.

mate, a fungitoxic product of benomyl, in the seeds at the time when protection is needed most. Seed germination is higher and occurrence of *Phomopsis* and total fungi lower when fungicides are applied at mid-pod or early-pod plus mid-pod stages.[37] One benomyl (0.05%) spray before flowering and another during flowering are as effective as five sprays of benomyl after flowering to control of purple seed stain (*Cercospora kikuchii*) of soybean.[38] Fungi may gain entry to soybean seeds in unopened pods through cracks in pod walls, insect injuries, and/or by systemic infection.[39,40] Benomyl may persist in plant parts for up to 30 days after spraying.

Foliar fungicides are used to control anthracnose, Cercospora leaf blight, frogeye leaf spot, *Phomopsis* seed decay, purple seed stain, Septoria brown spot, soybean rust, stem blight, and stem canker on soybeans.[17] All except the soybean rust fungus are seedborne. Growers in tropical and subtropical areas and seed producers in all regions may benefit from the proper application of foliar fungicides. Benefits from fungicide application include higher yields, better seed germination, less fungal seed infection, larger seeds, and seeds that appear cleaner. If seedborne fungi are controlled effectively by foliar fungicides, seed dressing treatment may not be necessary. Checklists have been developed at the University of Illinois at Urbana-Champaign (Table 19) and the University of Kentucky, Lexington (Table 20) to help determine conditions when these diseases are likely to be severe enough to apply a fungicide(s).[17] At present, two fungicides — benomyl and thiabendazole — are approved for use as foliar fungicides on soybean in the U.S. To control both foliar and late-season diseases and to achieve the highest yield and seed quality, fungicide protection should be provided from bloom (R_2 growth stage) to harvest (R_8 growth stage) (Figure 52; Table 18). Growers who wish to increase yield should begin fungicide application during pod-fill (R_6 growth stage) while seedsmen interested in seed quality should time applications to ensure that pods are protected until harvest.[17] In 1978, approximately 3.5 million acres of soybeans were sprayed with fungicides in the U.S. to control soybean diseases and to reduce seedborne inoculum.[41] Spraying soybeans with fungicides is profitable particularly when *Phomopsis* seed decay is a problem.[42]

2. Wheat

Field studies in southeast England with winter wheat during 1973—75 show that a single benomyl or benomyl plus maneb spray applied between flag leaf and ear emergence protects the leaf and ear and results in less seedborne infection of *Septoria nodorum* and *S. tritici*.[43] Foliar application of captafol 1.21 kg active ingredient per hectare and mancozeb at 1.45 kg active ingredient per hectare reduced wheat seed infection by *S. nodorum*.[7] Wheat seed infection by *F. graminearum* was reduced by spraying plants with benomyl + mancozeb, or benomyl or methylbenzimidazole carbamate alone.[44]

3. Maize

Foliar sprays of triphenyltin acetate and copper oxychloride reduced seedborne *F. moniliforme* and *Curvularia pallescens* and white streak on maize kernels. The percentage of streaking in seeds was 39.2 from unsprayed plots; in those sprayed with Bordeaux mixture, 36.0; with fentin hydroxide, 31.2; copper oxychloride plus zineb, 27.6; triphenyltin chloride, 26.8; mancozeb, 17.6; triphenyltin acetate, 11.8; and copper oxychloride, 9.[45]

4. Rice

Spraying rice with the following fungicides reduced seed infection by *Alternaria longissima*, *A. padwickii*, *C. lunata*, *Drechslera oryzae*, *Phoma* spp., and *Sclerotium* spp.: triphenyltin chloride, fentin hydroxide, IBP (Kitazin), zineb, and mancozeb.[46] The in-

Table 19
A POINT SYSTEM FOR DETERMINING WHETHER TO APPLY FOLIAR FUNGICIDES TO SOYBEANS (*Glycine max*) FOR ILLINOIS

Condition[a]	Point value[b]
Rainfall, dew, and humidity up to R_1 to R_2 (early bloom) and R_3 (pod set)	0
Normal	2
Above normal	4
Cropping history	
Soybeans grown previous year	2—3
Tillage practices	
Chisel-plow, disk, or no-till	1
Disease symptoms	
Pycnidia (black specks) visible on fallen petioles or Septoria brown spot obvious on lower leaves	2—3
Cultivar selection	
Early-maturing cultivar	1—2
Full-season cultivar[c]	0
End use	
Soybeans to be used or sold for seed	3
Yield potential	
Better than 2354 kg/ha (35 bu/acre)	2
Below 2354 kg/ha (35 bu/acre)	0
Seed quality (warm germination test)	
Less than 85% germination	1
85% or more germination	0
Other conditions that favor disease development	
Weather forecasts indicate a 30-day period of above-normal rainfall, field has a history of disease, etc.	1—3

[a] Observations should be made at the R_3 growth stage (first pod set).
[b] If a condition does not apply, the point value is zero. Add the point values for each applicable condition. If the point value is 12 or more for seed-production fields and 15 or more for grain-production fields, foliar fungicides will probably increase yield and seed quality.
[c] Foliar fungicides should not be used on full-season cultivars.

After Shurtleff, M. C., Jacobsen, B. J., and Sinclair, J. B., Pod and Stem Blight of Soybean, Report on Plant Diseases No. 509 (revised), Department of Plant Pathology, University of Illinois at Urbana-Champaign, 1980.

cidence of *A. padwickii* and *D. oryzae* was lowest in rice seeds when chlorothalonil (1.5 kg/ha), mancozeb (5 kg/ha), carboxin (2.51 *l*/ha) and sisthane (0.64 *l*/ha) were applied before the dough stage. Two applications of chlorothalonil or metam-sodium controlled *D. oryzae*. All the fungicides controlled *A. padwickii*.[47]

5. Sorghum

Weekly applications of benomyl + captan (0.5 kg + 0.5 kg/ha) from boot stage to

Table 20
A POINT SYSTEM FOR
DETERMINING WHETHER TO APPLY
FOLIAR FUNGICIDES TO SOYBEANS
(*Glycine max*) FOR KENTUCKY

Risk factor	Point value[a]
Cropping history	
Soybeans grown previous 2 or more years	3
Soybeans grown previous year	2
Soybeans not grown previous year	0
Cultivar selection	
Early-season cultivar	3
Mid-season cultivar	2
Late-season cultivar[b]	0
Planting date	
Before May 20	3
Between May 20 & June 20	2
After June 20	0
Rainfall[c]	
Below normal	0
Near normal (± 2 cm [0.5 in.])	2
Above normal	4

[a] If the point total is 11 or more, a fungicide should be applied; if it is 9 or 10, a fungicide may be beneficial; if it is 8 or less, fungicides should not be applied.

[b] Foliar fungicides should not be used on late-season cultivars.

[c] Based on recorded precipitation and predicted rainfall during seed development and maturation from growth stage R_2 to R_7.

After Stuckey, R. E., Jacques, R. M., TeKrony, D. M., and Egil, D. B., Foliar Fungicides Can Improve Soybean Seed Quality, Kentucky Seed Improvement Association, Lexington, Kentucky, 1981.

physiologic maturity significantly reduced *C. lunata* and completely controlled *F. moniliforme* in sorghum seeds. Fungicide application increased seed yield, 1000-seed weight, and germination.[48] A spray of mancozeb (0.2%) is more effective in reducing sorghum ergot than one of captafol (0.2%) and ziram (0.2%).[49]

6. Brassica spp.

Three sprays of iprodione (50% active ingredient) at 0.5 to 1 kg active ingredient per hectare on *Brassica oleracea* (cabbage) seed crops at 3-week intervals from the young green-pod stage until cutting controlled pod and seed infection by *A. brassicicola*. Only a few seeds were infected, seed yield increased, and germination was improved.[50]

7. Bean

No seed infection was recorded in beans by *Pseudomonas syringae* pv. *phaseolicola* and *Xanthomonas campestris* pv. *phaseoli* from plants sprayed with copper oxide every 7 days, and 0.28% infection by *P. syringae* pv. *phaseolicola* and none by *X. campestris* pv. *phaseoli* was recorded from plants sprayed at 14-day intervals.[51] Chlorothanlonil (0.1 % active ingredient) sprays two to three times prevent *Ascochyta fabae* infection in broad beans (*Vicia faba*).[52]

8. Pigeonpea
Pigeonpea seeds from plants sprayed two to four times with benomyl (2.2 kg/ha) produced fewer seedborne fungal infections.[53]

9. Okra
Injury by spotted bollworm (*Earias fabia* and *E. insulana*) increases the incidence of *Aspergillus flavus, F. moniliforme, F. oxysporum, F. semitectum,* and *Macrophomina phaseolina* in okra seeds. A preharvest spray with the insecticide monocrotophos (0.1%) reduces the incidence of these fungi. A carbendazim (0.1%) + monocrotophos spray almost completely controlled these fungi.[54]

N. Insect Control
Insect control reduces stress, reduces injuries where pathogens may enter, and may reduce virus spread. A regular insecticide spray schedule to control insect infestation can be useful in maintaining low infection levels in seeds.

O. Weed Control
Weed control is important because weeds compete for space, nutrients, and water, and can be hosts to pathogens. Plants under stress from weed competition are more susceptible to most pathogens. Hand hoeing and mechanical and chemical weed control methods vary from one region to another. Herbicides must be used with care to avoid plant damage. Evidence is increasing that many weeds in and around soybean fields are susceptible to soybean pathogens, including seedborne pathogens such as *Colletotrichum* and *Phomopsis* spp., and SMV and TRSV.[17]

There was a signficant correlation between development of weeds (*Alternanthaera ficoidea, Braccharis* sp., and *Commelina diffusa*) and the occurrence of *Colletotrichum truncatum, F. semitectum,* and *P. sojae* in soybean seeds.[55] Weeds, such as *Abutilon theophrasti, Amaranthus spinosus, Leonotis nepetaefolia,* and *Leonurus sibiricus,* can act as alternative hosts or provide a microclimate of prolonged humidity favoring seed infection.[56,57]

P. Harvesting
Seeds should be harvested immediately when mature. The longer the seeds remain in the field after maturity, the greater the chance for invasion by pathogenic bacteria and fungi, especially under warm, moist conditions. Harvesting and threshing the plots with a low seed infection should be done prior to those plots having a high seed infection to avoid mixing healthy and diseased seeds or contaminating healthy seed lots. Clean cottonseed can be contaminated by *X. campestris* pv. *malvacearum* infected dry refuse in or on harvesting equipment, trucks, or trailers used to take picked cotton to gins.[58]

IV. SEED TREATMENT

Seed treatment is a biological, chemical, mechanical, or physical process designed to mitigate externally or internally seed- or soilborne microorganisms, resulting in the emergence of a healthy seedling and subsequently a healthy plant. Seeds may be treated to promote good seedling establishment, to minimize yield loss or to maintain and improve quality, and to avoid further spread of pathogens.

A. Biological Control

Biological control is the reduction of inoculum intensity or disease-producing activities of a pathogen or parasite in its active or dormant state, by one or more organisms, accomplished naturally or through manipulation of the environment, host, or antagonist, or by mass introduction of one or more antagonists.[59] *Trichoderma viride* was the first fungus demonstrated as an antagonist for control of soilborne pathogens, such as *Rhizoctonia solani*.[60] Biological control of soilborne fungi has been studied extensively but demonstrated in few cases.[60] Results from studies on the control of seedborne pathogens and better plant stands through the application of antagonistic fungi and bacteria to seeds have been inconsistent.

Antagonists applied commonly to seeds are *Bacillus subtilis, Chaetomium* spp., *Penicillium oxalicum*, and *Trichoderma* spp. Their application to seeds reduces seedborne fungi and results in vigorous seedlings.

Pea seeds treated with *P. oxalicum* produced plant stands, vine, and pod weights equal to those produced from captan-treated seeds and were significantly better than those produced from untreated seeds.[61] The number of spores per seed necessary to protect a plant varies with inoculum concentration of the pathogen in the soil. For treating pea seeds, approximately 6×10^6 spores of *P. oxalicum* per seed provides protection for pre- and postemergence damping-off similar to that of captan.[61] Spores of *P. oxalicum* germinate and produce hyphae which grow between root hairs and on the root surface.[62]

T. hamatum applied to pea or radish seeds controls seed rot in soil infested with *Pythium* sp. or *R. solani* at soil temperatures between 17 to 34°C. Seeds are treated with a conidial suspension at a concentration equal to or greater than $10^6/m\ell$. Seed treatment with *Rhizobium* and *T. hamatum* had no adverse effect on the nodulating activity of the former or the protective ability of the latter.[63] *B. subtilis* and *C. globosum* applied to maize seeds controlled seedling blight due to *Fusarium moniliforme* and *F. roseum* f. sp. *cerealis* "*graminearum*". The treatment increases emergence, root vigor, plant fresh weight, root dry weight, and seedling stand. Treating kernels with either microorganism gave control equal to treatment with captan or thiram when the soil temperature was under 20°C.[64,65] A *Chaetomium* sp. isolate from oat seeds protects oat seedlings from infection by seedborne *Drechslera victoriae*.[66] Certain isolates of *C. cochlioides* and *C. globosum* controlled Fusarium blight (*F. nivale*) in oats as effectively as an organic mercury seed treatment.[67] Seed treatment of sweet corn and wheat with *T. harzianum* is an effective component of crop management[68] and when used on snap bean reduces *Rhizoctonia solani* damping-off in acid soils.[69]

Treating flax seeds with *B. fluorescens* and *B. mesentericus* reduces seedling disease caused by *Colletotrichum* and *Fusarium*.[70] *B. subtilis* and *Streptomyces* sp. applied to wheat seeds reduce the effects of *R. solani* and stimulate seedling growth.[71] Five of nine *B. subtilis* isolates applied to maize seeds and sown in *F. roseum* "*graminearum*"-infested soil in the greenhouse at 18°C resulted in significantly higher stands. The use of antagonistic microorganisms may be as effective as captan in reducing the incidence of seedborne infection.[72] Transmission of *C. lini* and *Polyspora lini* in flax seeds is lower in natural than sterilized soil due to microorganism activity. A bacteria was found which caused conidial disintegration of *C. lini* in soil.[73]

Bacteriophages have been used to control seedborne bacteria. Maize seeds treated with a phage and subsequently inoculated with *Erwinia stewartii* developed 1.4% compared to 18% infected plants from untreated seeds.[74,75]

TMV in tomato seeds is partially inactivated or inhibited during germination by normal physiological processes or through soil microflora. Five days after sowing, levels of infected seeds dropped from 94 to 44%.[76]

The use of biological control agents has not been found practical under field conditions. Results have been inconsistent because of variables such as types of coating materials used as carriers of antagonists, moisture and temperature effects, and duration of storage after treatment.

1. Testing Antagonists for Seed Treatment

Seed treatment has been used in various crops, generally using antibiotic-producing antagonists, for control of pre- and postemergence damping-off and seedling root rot diseases.[72,77-81] Antagonists are grown in liquid or agar media. When bacteria are grown on a liquid medium, they are agitated during incubation, while Actinomycetes or fungi are cultured without agitation. Bacterial cells are collected by centrifugation and resuspended in physiological saline or used directly. Actinomycete and fungal cultures are filtered and the mat is blended in sterile water and used as a seed treatment. When the antagonists are suspended in water, the seeds are immersed in the suspension for 15 min to about 1 hr, then air dried at moderate temperatures and planted either immediately or stored in a refrigerator in paper bags. The dry treatment generally is used for Actinomycete and fungal spores. A weighed quantity of spores is added to seeds slightly moistened by water or an adhesive agent such as 4% carboxyl-methyl-cellulose or gum arabic. The seeds and spores are mixed so that all the seeds are covered by the spores. Rolling the moist seeds over the heavily sporulating colony of the antagonist on agar media also is useful. Rolling also is used for inoculation with bacterial antagonists. The bacterium is grown on an agar medium and after 48 hr the colonies are scraped off and mixed with a small amount of distilled water in a culture plate. The seeds are rolled over the viscous fluid until most of the liquid has been absorbed, then air dried at room temperature.

To determine the efficacy of the antagonists as seed treatment, one or two standard seed-treatment fungicides should be included in the tests for comparison. Initial testing is done on blotter paper or sterile sand in pathogen-infested soil of different types and finally in the field.

B. Chemical Method

Application of chemicals to seed is the safest, cheapest, and most effective means of controlling most seedborne pathogens. Fungicidal seed treatment may kill or inhibit seedborne pathogens and may form a protective zone around seeds that can reduce seed decay and seedling blight caused by soilborne pathogens, resulting in healthy and vigorous seedlings. The use of fungicides as seed treatments is the most widely followed disease control practice used in all crops.[82,83] The first mention of seed treatment for control of plant diseases is by Caius Plinius Secundus, Pliny the Elder, in an encyclopedia of Roman agriculture titled, *Historia Naturalis* (*circa* 23—79 A.D.). It refers to corn mildew (wheat smut) control by ''...steeping of the seed for planting in wine or mixing bruised cypress leaves with it is to be recommended.''[83] More than 1000 years ago the importance of good seeds was realized in India in Surpala's *Vrksayurveda* (800 A.D.): "A seed which has been steeped in milk and rubbed well in cowdung and after drying, again rubbed repeatedly in honey and the powder of vidanga (*Embelia ribes*) grows without fail."[82]

It was an accident which lead to the general use of seed treatment for control of plant diseases. In 1670, a sailing vessel loaded with wheat encountered a storm in the Bristol Channel, England. The wheat, saturated with salt water, was grounded. Farmers along the coast salvaged some grains and used them as planting seeds. The grain from such plants was free of stinking smut whereas grain produced from native seeds was infested. This led to the soaking of wheat seeds in salt brine in England during the 17th and 18th century for control of stinking smut of wheat. In 1637, Remnant reported the use of common salt as a seed treatment of wheat for control of stinking smut of wheat. In 1733, Jethro Tull published the method of wheat seed treatment with salt brine in detail for control of stinking smut. Later, Schulthess, in 1760, suggested the use of a seed treatment for control of stinking smut of wheat. The use of copper compounds was supported by Tillet in 1755 and Prevost in 1807.[82,83]

In 1913, Reihm suggested the use of organic mercurial fungicides as a seed treatment for control of stinking smut of wheat. Following World War II, organic mercurials were introduced commercially for treatment of vegetables and small-grain seeds. However, mercury poisoning of a family in New Mexico as a result of consuming a pig fed with mercury-treated wheat seeds in 1969—70 resulted in a ban on the use of mercury in the U.S. However, mercury compounds are used as seed treatments.[82,83]

Two quinone compounds were introduced as seed protectants in the 1940s. The first organic seed protectant, chloranil, was introduced by Cunningham and Sharvelle in 1940. Felix and Terhorst introduced dichlone in 1943, a dithiocarbamate seed protectant; thiram, introduced in 1941 by Harrington for control of turf diseases, was recommended as a seed protectant in 1951. In 1952, a heterocyclic compound, captan, was introduced as the miracle seed protectant fungicide. The discovery of the systemic fungicide carboxin by Uniroyal Chemical Co. in 1966 revolutionized the control of plant diseases by seed treatment. Carboxin replaced all nonchemical methods for the control of loose smut of wheat. This led subsequently to the introduction of large number of systemic fungicides. Benomyl was introduced by E. I. duPont deNemours & Co., Inc. in 1968. Other compounds such as fenfuram, metalaxyl, triadimexor, and others were introduced later by other companies. Antibiotics, such as streptomycin and tetracyclines, also are used for the control of seedborne bacteria.[82,83]

The chemical nature and trade names of fungicides marketed in different countries for seed treatment are given in Table 21.[82,83]

1. Formulations

Seed treatment chemicals are available in different formulations.[82,83]

Wettable powders (WP) — This type of formulation usually contains a wetting agent. It aids in dispersal of the chemical in water. Water-dispersible formulations are preferred for slurry or mist seed treatment. Most fungicides are available as water-dispersible powders.

Dusts — Dust formulations are dry powders and are used for slurry or mist treatments.

Slurries or suspensions — The chemical is mixed with a liquid in high concentrations and diluted at the time of seed treatment.

2. Categories of Chemical Seed Treatment

Control through chemical seed treatment may be classified in the following categories.[84]

a. Seed Disinfection

This refers to the control of inoculum established within the seed or seed coat tissues. Such pathogens are controlled by thermotherapy (hot air, oil, or water treatment), or by systemic or mercurial fungicides which are absorbed, penetrate, or diffuse inside the seed. Seed disinfection chemicals are specific and are applied after an assessment of the seedborne inoculum.

b. Seed Disinfestation

This refers to the control of the pathogens which are externally or passively present on the seed surface. It is easier to control such pathogens by fungicide seed treatment than those microorganisms that infect seeds.

Table 21

FUNGICIDES USED AS SEED TREATMENTS TO CONTROL SEEDBORNE PATHOGENS[82,83]

Common name	Trade name	Active ingredient
Nonsystemic fungicides		
Sulfur fungicides		
Inorganic sulfur		
Elemental sulfur		Sulfur
Organic sulfur (carbamates)		
Ferbam	Hexaferb, Coronet, Ferberk, Fermate, Fermocide, Ferradow, Karbamblack	Ferric dimethyl dithiocarbamate
Maneb/mancozeb	Dithane M-22, Manzate, MEB, MnEBD	Manganese ethylenebisdithiocarbamate
Thiram	Arasan, Hexathir, Nomersan, Fermide, Fernacol, Thiride, TMTD, Thylate, Tersan, Fernasan, Spottrete	Tetramethyl thiuram disulfide
Zineb	Dithane Z-78, Hexathane, Lonacol, Parzate C, duPont Fungicide A	Zinc ethylene bisdithiocarbamate
Mercury fungicides		
	Agrosan GN	1% Hg (0.9% phenyl mercury acetate + 0.1% ethyl mercury chloride)
	Agrox (U.S.)	Phenyl mercury urea
	Aretan 6, Agallol, Ceresan wet (India)	Methoxy ethyl mercury chloride
	Emisan 6, Tayssato	Ethyl mercury chloride
	Ceresan (U.S.)	Methoxyethyl mercury silicate
	Ceresan (Germany)	N-(ethylmercuri)-p-toluene-sulfonanilide
	Ceresan M (U.S.)	Phenyl mercuric acetate
	Ceresan dry (India), Ceresan Universal, Agrosan, Dynacide, Gallotox, Leytosan	Ethyl mercurithiosalicylic acid (sodium salt)
	Elicide (U.S.)	N-(ethylmercuri)-1,4,5,6,7,7-hexachloro-bicyclo (2,2,1) hept-5-ene-2,3,dicarboximide
	EMMI (U.S.)	Ethyl mercury chloride
	Granosan	Methylmercury 8-hydroxyquinolate chloromethoxypropyl-mercuric acetate
	LM seed protectant Quicksan (U.S.)	

Quinone fungicides

Chloranil		2,3,5,6-Tetrachloro-1,4-benzoquinone
Dichlone		2,3-Dichloro-1,4-naphthoquinone

Heterocyclic nitrogenous compounds

Captan	Captan, Esso fungicide 406, Orthocide, Vancide 89, Captane, Vondcaptan	N-(Trichloromethylthio)-4-cyclohexene-1,2-dicarboximide
Captafol	Difolatan, Difosan, Sanspor, Sulfonimide	cis N-(1,1,2,2-tetrachloroethylthio)-4-cyclohexene-1,2-dicarboximide

Systemic fungicides

Triadimenol	Baytan Universal	(Combination product — with fuberidazole & imazalil)
Triadimenol	Baytan 15SD	1-(4-Chlorophenoxy)-3,3-dimethyl-1,(1,2,4-triazole-1-yl) butanone
Triadimefon	Bayleton 25 WP	1-(4-Chlorophenoxy)-3,3-dimethyl 1-1-(1-H-1,2, 4-triazol-1-yl)-2-butanone
Benomyl	Benlate, Grex, Tersan 1991, Ultra sofril	Methyl-(1-butylcarbamoyl)-2-benzimidazole carbamate
Carbendazim	BAS 3460, Bavistin, Derosal, MBC	Methyl-2 benzimidazole carbamate
Carboxin	DMOC, Vitavax	5,6-Dihydro-2-methyl-1, 4-oxathiin-3-carboxanilide
Fenfuram	Panoram	N-phenyl-2-methylfuran-3-carboxamide
Furcarbanil	BAS 3191F	2,5-dimethyl-3-furanilide
Metalaxyl	Apron 35SD	Methyl DL- N-(2,6-dimethyl phenyl)- N-2' methoxyacetyl)-alaninate (35% metalaxyl)
Oxycarboxin	DCMOD, Plantvax	2,3-Dihydro-5-carboxanilido-6-methyl-1,4-oxathiin-4,4-dioxide
Pyracarbolid	Sicarol	2-Methyl-5,6-dihydro-4H pyran-3-carboxalic acid anilide
Tridemorph	Calixin M	11% tridemorph + 36% maneb (11% 2,6-dimethyl-4-tridecyl morpholine + 36% maneb)
Quintozene + ethazol	Terracoat L-205	Pentachloronitrobenzene(23.2% + Terrazole 5.8%) PCNB + 5-ethoxy-3 (trichloromehyl)-1,2,4-thiadiazole

	New Improved Ceresan (U.S.)	Ethyl mercury phosphate
	Panogen (U.S.)	Methyl mercuric dicyandiamide
	Semesan (U.S.)	Hydroxy mercuric chlorophenol
	Spergon	
	Phygon, phygon XL	

Table 21 (continued)

FUNGICIDES USED AS SEED TREATMENTS TO CONTROL SEEDBORNE PATHOGENS[82,83]

Common name	Trade name	Active ingredient
Carboxin + thiram	Vitavax 200	(37.5% carboxin + 37.5% thiram)
Antibiotics		
Cycloheximide	Actidione	β-(2-(3,5-Dimethyl-2-oxocyclohexyl)-2-hydroxyethyl) glutarimide
Streptomycin	Agrimycin-17, Agri-strep, Rimocidin, Chem-form	2,4-Diguanidino-3,56-trihydroxycyclohexyl-5-deoxy-2-O (2-deoxy-23 methylamino α-glucopyranosyl)-3-formyl pentantofuranoside
Miscellaneous fungicides		
PCNB	Quintozene	Pentachloronitrobenzene
Lesan	Dexon	Sodium-P-(dimethyl amino benzene) diazosulfonate

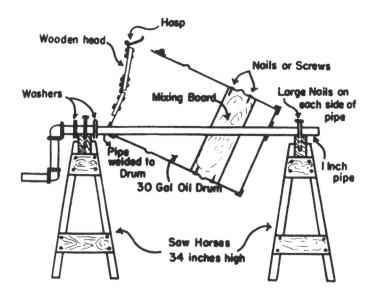

FIGURE 53. An oil-drum or rotary seed treater. (From Sharvelle, E. G., *Plant Disease Control*, AVI Publishing, Westport, Conn., 1979, 203. With permission.)

c. Seed Protection

This method involves seed treatment with a suitable fungicide which protects the seed and seedling from seed- and soilborne microflora. Many soilborne fungi are facultative parasites which under suitable environmental conditions cause seed rot and seedling blight. Ideally, all seeds planted under favorable environmental conditions should have enough vigor to germinate and emerge. However, poor emergence often is a problem under field conditions. The lack of emergence or seed rot may be attributed to genetic, physiologic, mechanic, or pathologic factors. There are situations in which good-quality seeds germinate and emerge poorly because of the association with seed rot caused by certain pathogens. This type of seed rot is common with seeds of most crop plants, especially legumes and vegetables. There may be both pre- and postemergence damping-off.

Both seed- and soilborne microorganisms are responsible for causing seed decay. Many of the pathogenic seedborne fungi are species of *Alternaria, Aspergillus, Botrytis, Cephalosporium, Cercospora, Colletotrichum, Curvularia, Cladosporium, Drechslera, Fusarium, Phoma, Phomopsis,* and *Verticillium.* Common soilborne fungi associated with seed rots are species of *Botrytis, Fusarium, Phytophthora, Pythium,* and *Rhizoctonia.* Environmental conditions, such as an excess of moisture, heavy soils, careless handling, and presence of weeds also favor seed rots. The application of a general seed protectant to seed helps in producing better emergence and vigorous seedlings. Seed-protectant chemicals differ from crop to crop and from region to region.[85]

3. Method of Treatment

Dry seed treatment — Seeds can be treated in small amounts in seed treaters (Figures 53 and 54). However, some fungicide is lost to the sides and base of the treater and the chemical may not stick to the seed surface. Generally, this method is used by farmers who treat their own seed. Dry seed treatment generally is recommended for treating legume seeds, especially soybean, where the seeds are apt to be damaged during seed processing, and in cases when the slurry or dip methods are not applicable due to

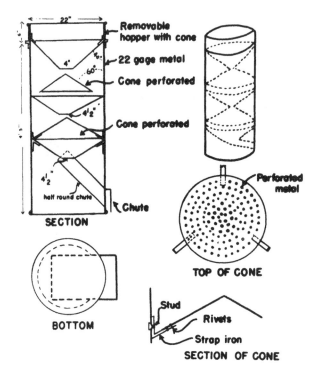

FIGURE 54.   A gravity-type seed treater. (From Sharvelle, E. G., *Plant Disease Control*, AVI Publishing, Westport, Conn., 1979, 210. With permission.)

absorption of the water by the seeds, or for other reasons where the dry method is more reliable.

Seed dip — This method involves dipping seeds into a fungicide solution for a time depending upon the fungicide and type of seed to be treated and is used when the seed coat is thick. It helps in the better absorption of the chemical. The seed-dip method generally is used just prior to planting. Streptomycin is used in a soak treatment for eradication of deep-seated bacterial infections in beet (*Beta vulgaris*)[86] and bean (*Phaseolus vulgaris*) seeds.[87] Immersion of crucifer seeds into an antibiotic solution (tetracyclines <3000 μg/mℓ for 30 min) reduces *Xanthomonas campestris* pv. *campestris* on naturally infected seed to 1% and on inoculated seed to <8%.[88]

Planter or hopper-box treatment — In this method the required amount of fungicide is mixed with the planting seeds in the hopper-box of a seed-drill just before planting. The chemical is not wasted and only the amount of seed required for planting is treated.

Pelleting — Fungicides can be pelleted on seeds. Seeds of pine[89], onion,[90] and sugar beet[91] have been pelleted with fungicides to control of a variety of seedborne fungi. A limitation of this method is the slow absorption of fungicide by seeds. The method is used only if other methods are unsuitable.

Fumigation — Eradication of seedborne pathogens with fumigants requires further seed treatment with a seed protectant against soilborne microflora. Fumigation of high-quality soybean seeds with formaldehyde for 2 hr, propylene oxide for 9 hr, or hydrazoic acid for 53 hr results in 83, 83, and 75% bacteria-free seed, respectively, and eliminates most surface fungi. Formaldehyde or propylene oxide fumigation does not decrease seed vigor or increase seed leachate conductivity.[92] Ethylene oxide gas controls *F. avenaceum* in clover, *Pseudomonas syringae* pv. *phaseolicola* in bean, and *P. syr-*

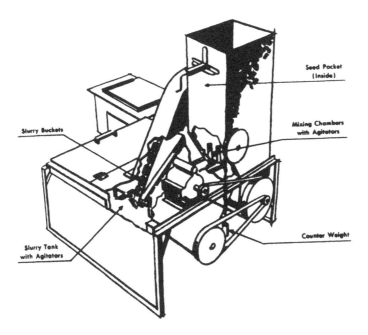

FIGURE 55. A slurry seed treater. (From Sharvelle, E. G., *Plant Disease Control*, AVI Publishing, Westport, Conn., 1979, 212. With permission.)

ingae pv. *glycinea* and *X. campestris* pv. *phaseoli* var. *sojense* in soybean seeds with vacuum fumigation. Seeds are fumigated under a sustained vacuum in a heated 17^3-m chamber. The gas is introduced as a 90% ethylene oxide plus 10% carbon dioxide mixture. Seeds are kept in polyethlyene bags which are opened prior to fumigation and resealed immediately after treatment.[93] *Allium cepa* (onion) seed can be disinfected of the nematode *Ditylenchus dipsaci* by fumigating with 40 oz of methyl bromide per 1000 ft^3 for 24 hr. This also kills nematodes in stem pieces. There is no deleterious effects on seed viability.[94] Satisfactory control of *Anguinia agrostis* (nematode) in bent grass (*Agrostis tenuis*) seeds is obtained by fumigating infested seeds of about 12% moisture with methyl bromide.[95]

Slurry treatment — In slurry treatments (Figure 55), water-dispersible fungicide formulations are mixed in water so that a slurry formation results, which is applied to seeds. The slurry treatment is used primarily to treat seeds after harvest and before seed processing. The seeds are coated with the fungicide and it sticks on the surface of the seed. It is easy to differentiate between treated and untreated seed lots. Most fungicides used for seed treatment can be treated using a slurry method.[82,83]

Mist treatment — In this technique, a fungicide suspension is broken into fine droplets during the treatment process. This is considered to be the most effective means of treating seeds. The seeds are coated uniformly with the fungicide. Mistomatic treaters currently are being used by most of the seed-processing agencies throughout the world.[82,83]

4. Application

Seed treatment with chemicals has been used for the control of seedborne bacteria, fungi, nematodes, and viruses. Important seedborne bacteria and fungi controlled through seed treatment are given in Table 22.

Control of seedborne viruses through chemical seed treatment has been attempted. Treatment of pepper (*Capsicum annuum*) seeds with calcium or sodium hypochlorite,

Table 22

CONTROL OF IMPORTANT SEEDBORNE BACTERIAL AND FUNGAL PATHOGENS THROUGH CHEMICAL SEED TREATMENT

Crop	Disease	Pathogen	Chemical	Dosage[a]	Ref.
Allium cepa (onion)	Gray mold	*Botrytis allii*	Benomyl	0.1%	96
	Smut	*Urocystis cepulae*	Pelleting with captan or thiram		83
	Seed protectant	Several fungi	Ferbam	0.15 to 0.2%	97
			Seed pelleting with captan 50 WP	303 g/4.5 kg	83
			Thiram 75 WP	303 g/4.5 kg	83
			Captan 50 WP	156 g/45 kg	83
			Thiram 75 WP	156 g/45 kg	83
Apium graveolens (celeriac, celery)	Early blight	*Cercospora apii*	Thiram	—	83
Arachis hypogaea (groundnut, peanut)	Seed protectant	Several fungi	Captan + quintozene (3:1)	0.3%	98
Avena sativa (oats)	Loose smut	*Ustilago avenae*	Furcarbanil	1%	99
			1% organomercurial	0.2 to 1.5%	100
	Covered smut	*U. hordei*	Carboxin	0.2%	101
			Pyracarbolid	0.1%	102
			1% organomercurial	0.2 to 1.5%	103
	Seed protectant	Several fungi	Captan 50 WP	57 to 85 g/bu	83
			Mancozeb 80 WP	43 g/bu	83
Beta vulgaris (beet)	Blackleg	*Phoma betae*	Thiram	0.25%	104
	Seed protectant	Several fungi	Bavistin 50 WP	0.25% soak	105
			Benomyl	0.25% soak	105
			Captan 83 WP	0.25% soak	105
			Ceresan dry (1% Hg)	0.25%	106
			Dexon 50 WP	85 g/45 kg	83

Host	Disease	Pathogen	Treatment	Dosage	Ref.
Brassica spp. (crucifers)	Leafspot	*Alternaria brassicicola*	Quintozene	0.25%	106
		A. brassicae	Thiram 50 WP	99 g/45 kg	83
	Black leg	*P. lingam*	Iprodione	2.5 g ai/kg	107
			Thiram	0.15 to 0.25%	108
			Benomyl	0.25%	5
	Black rot	*Xanthomonas campestris* pv. *campestris*	Agrimycin 100	0.01%	109
			+ Bangtan	0.2%	
			Streptocycline	0.01%	
			+ Bangtan	0.20%	109
Capsicum frutescens (pepper)	Seed protectant	Several fungi	Thiram 75 WP	85 g/45 kg	83
			Captan 50 WP	51 g/45 kg	83
	Damping-off, anthracnose	*Colletotrichum capsici*	Thiram	0.2%	110
	Wilt	*Fusarium oxysporum*	Bavistin, carboxin	0.2%	111
			Captan 83 WP	0.2%	112
	Seed protectant	Several fungi	Thiram 75 WP	156 g/45 kg	83
			Captan 50 WP	128 g/45 kg	83
Cicer arietinum (chickpea)	Chickpea blight	*Ascochyta rabiei*	Calixin M	0.3%	113
			12 hr immersion in pimaracin	150 µg/mℓ	114
			Granosan	—	115
			Thiram	0.3%	116
			Benomyl	0.3%	117
			Thiabendazole	0.3%	117
Coriandrum sativum (coriander)	Stem gall	*Protomyces macrosporus*	Thiram	0.25%	118
Corchorus capsularis (jute)	Stem rot	*Macrophomina phaseolina*	Captan	0.6%	119
Cucurbita spp. (cucurbits)	Seed protectant	Several fungi	Thiram 75 WP	0.25%	120
			1% organomercurial	0.25%	120
	Seed protectant	Several fungi	Captan 50 WP	2.5 tsp/1.9 kg	83
			Dexon 70 WP	1.25 tsp/1.9 kg	83
			Thiram 75 WP	3.75 tsp/1.9 kg	83

Table 22 (continued)
CONTROL OF IMPORTANT SEEDBORNE BACTERIAL AND FUNGAL PATHOGENS THROUGH CHEMICAL SEED TREATMENT

Crop	Disease	Pathogen	Chemical	Dosage[a]	Ref.
Daucus carota (carrot)	Seed protectant	Several fungi	Thiram 50 WP	85 to 113 g/45 kg	83
			Captan 50 WP	112 g/45 kg	83
Glycine max (soyabean, soybean)	Purple stain	*Cercospora kikuchii*	Benomyl + thiram (1:1)	0.4%	38
	Downy mildew	*Peronospora manshurica*	Spergon	0.25%	121
	Seed protectant	Several fungi	Thiram	0.25%	121
			Captan 83 WP	0.5%	122
			Thiram 75 WP	0.5%	123
Gossypium spp. (cotton)	Seedling blight	*Rhizoctonia solani*	Mancozeb	0.25%	124
	Blackarm	*X. campestris* pv. *malvacearum*	Thiram	0.25%	125
	Seed protectant	Several fungi	Captafol	0.25%	126
Helianthus annus (sunflower)	Downy mildew	*Plasmopara halstedii*	Apron 35 SD	0.6%	127
	Soft rot, blight	*Sclerotinia sclerotiorum*	7.7% ethyl Hg-*p*-toluene sulfanilide	—	128
	Seed protectant	Several fungi	1% organomercurial	0.25%	129
			Captan 75 WP	0.25%	130
			Mancozeb 80 WP	0.25%	130
Hordeum vulgare (barley)	Leaf stripe	*Drechslera graminea*	Carboxin	0.15%	131
	Covered smut	*U. hordei*	1% organomercurial	0.1 to 0.4%	132
			Carboxin	0.2%	133
			PMA/Ceresan M	0.2%	134
	Loose smut	*U. nuda*	Carboxin	0.25%	135
			Pyracarbolid	25 g ai/100 kg	102
			Furcarbanil	0.2%	136
Lactuca sativa (lettuce)	Seed protectant	Several fungi	Captan 80 WP	57 to 85 g/bu	83
	Seed protectant	Several fungi	Thiram 75 WP	85 g/45 kg	83

Host	Disease	Pathogen	Chemical	Dosage	Ref.
Lens spp. (lentil)	Seed protectant	Several fungi	Captan	0.2 to 0.3%	137
Linum usitatis simum (flax, linseed)	Rust	*Melampsora lini*	1% organomercurial	0.25%	138
	Seed protectant	Several fungi	Thiram 75 WP	0.25%	139
Lycopersicon esculentum (tomato)	Blight	*Pseudomonas syringae* pv. *tomato*	Na hypochlorite dip	—	140
Nicotiana tabacum (tobacco)	Anthracnose	*Colletotrichum tabacum*	Thiram	0.12 to 0.24%	141
Oryza sativa (paddy, rice)	Stackburn	*Alternaria padwickii*	Mancozeb	0.25%	142
			1% organomercurial	0.3%	143
	Seedling blight	*D. oryzae*	Captan	0.25%	144
			1% organomercurial	0.2 to 0.25%	143
			Mancozeb	0.3%	145
	Foot rot	*Gibberella fujikuroi*	1% organomercurial	0.2 to 0.25%	146
			Benomyl	—	82
	White tip	*Aphelenchoides besseyi*	Furadan 75 PM	0.5%	147
	Seed protectant	Several fungi	1% organomercurial	0.25%	143
			Captan 75 WP	0.25%	144
Pennisetum typhoides (pearlmillet)	Downy mildew	*Sclerospora graminicola*	Ridomil	0.2% ai	148
Phaseolus vulgaris (bean)	Leaf and pod spot	*Ascochyta fabae*	Benlate T	0.2% ai soak for 8 hr	52
	Anthracnose	*C. truncatum*	Dichlone	0.25%	149
	Halo blight	*Pseudomonas syringae* pv. *phaseolicola*	Streptomycin	5 µg/ml for 1 hr	150
	Common blight	*X. campestris* pv. *phaseoli*	Kasugamycin	0.025%	151
			Streptomycin	5 µg/ml for 1 hr	151
	Seed protectant	Several fungi	Captan 50 WP	43 g/45 kg	83
			Chloranil	0.2 to 0.25%	152

Table 22 (continued)

CONTROL OF IMPORTANT SEEDBORNE BACTERIAL AND FUNGAL PATHOGENS THROUGH CHEMICAL SEED TREATMENT

Crop	Disease	Pathogen	Chemical	Dosage[a]	Ref.
Pisum sativum (pea)	Leaf and pod spot	*A. pinodes*	Dexon 70 WP	28 g/45 kg	83
			Thiram 75 WP	43 g/45 kg	83
	Downy mildew	*Peronospora pisi*	Granosan	0.3%	153
			Benomyl	0.4%	154
	Blight	*Pseudomonas syringae* pv. *pisi*	Apron 35 SD	—	155
			Streptomycin sulfate	2.5 g ai/kg	156
	Seed protectant	Several fungi	Chloranil	0.2%	157
			Captan 83 WP	0.25%	158
			Thiram 75 WP	57 g/45kg	83
			Dexon 70 WP	20 g/45 kg	83
Raphanus sativus (radish)	Leafspot	*Alternaria raphani*	Thiram	0.15 to 0.25%	159
Secale cereale (rye)	Seed protectant	Several fungi	Mancozeb 80 WP	43 g/bu	83
Sesamum indicum (sesame)	Root rot	*Macrophomina phaseolina*	Captan	0.3%	160
Solanum melongena (brinjal, eggplant)	Seed protectant	Several fungi	Thiram 75 WP	112 g/45 kg	83
Sorghum vulgare (milo, sorghum)	Downy mildew	*Sclerospora sorghi*	Apron 35 SD	0.2%	161
	Grain smut	*Sphacelotheca sorghi*	Elemental sulfur	0.5%	162
			Dichlone	0.12%	163
			Chloranil	0.2%	164
			1% organomercurial		165
			Thiram	0.2%	166

	Seed protectant	Several fungi	Thiram 75 WP	0.25%	167
			1% organomercurial	0.2%	165
Spinacia oleracea (spinach)	Seed protectant	Several fungi	Thiram 50 WP	85 to 113 g/45 kg	83
Triticum aestivum (wheat)	Stripe	*D. graminea*	Sisthane	—	168
			Benlate T	—	168
	Foot and root rots	*D. sativa*	1% organomercurial	0.25%	169
	Snow mold	*F. nivale*	Benomyl	0.1%	170
			Benomyl + radosan	(1 g + 2 g/kg)	170
	Glume blotch	*Septoria nodorum*	1% organomercurial	0.2%	171
	Hill bunt	*Tilletia caries*	Carboxin	0.25%	172
			Chloranil	0.1%	173
		T. foetida	1% organomercurial	0.25%	174
			Mancozeb	0.25%	175
	Flag smut	*Urocystis agropyri*	Oxycarboxin	0.3%	176
			Pyracarbolid	0.5%	82
	Loose smut	*Ustilago tritici*	Carboxin	0.2 to 0.25%	117,178
			Vitavax 200 (37.5% carboxin + 37.5% thiram)	0.4%	179
			Sicarol	0.2%	179
			Furcarbanil	0.2%	143
			Benomyl	0.25%	180
			Bavistin	0.25%	180
			Sisthane	—	168
			Benlate T	0.25%	168
			Baytan U	0.2%	168
Zea mays (corn, maize)	Seed protectant	Several fungi	Thiram 75 WP	0.25%	181
			Benomyl 50 WP	57 to 85 g/bu	181
			Captan 80 WP		83
			Mancozeb 80 WP	43 g/bu	83
	Anthracnose	*C. graminicola*	Terra-coat L-205	0.8%	182
	Leaf blight	*D. maydis*	Carboxin + thiram	0.25%	183
	Head smut	*Sphacelotheca reiliana*	Carboxin	0.25 to 0.4%	184

Table 22 (continued)
CONTROL OF IMPORTANT SEEDBORNE BACTERIAL AND FUNGAL PATHOGENS THROUGH CHEMICAL SEED TREATMENT

Crop	Disease	Pathogen	Chemical	Dosage[a]	Ref.
	Seed protectant	Several fungi	Captan 50 WP	14 to 57 g/45 kg	83
			Captafol	0.12%	185
			Chloranil	0.2%	186
			Dexon 70 WP	71 to 113 g/45 kg	83
			1% organomercurial	0.25%	187
			Thiram 50 WP	71 to 113 g/45 kg	83
Zinnia	Seedling blight	*A. zinniae*	1% organomercurial	0.25%	188
elegans			Thiram	0.25%	188
(zinnia)	Blight	*X. campestris* pv. *zinniae*	30 min soak in 10,500 µg/mℓ Na hypochlorite		189

[a] ai = active ingredient.

hydrochloric acid, or trisodium phosphate controls TMV in seedlings.[190] Treating to-matoiuaajs with 10% trisodium phosphate or extracting with concentrated hydro-chloric acid (10 mℓ/25 lb of fruit pulp) reduces TMV.[191,192] Soaking tomato seeds in 1% trisodium orthophosphate for 15 min and then in 0.525% sodium hypochlorite for 30 min reduces seed transmission of TMV.[193] Soaking cowpea seeds in malic hydrazide at 40, 100, and 400 μg/mℓ for 90 min, 2-thiouracil at 500 or 700 μg/mℓ for 60 min, naphathalene acetic acid (NAA) at 40 μg/mℓ for 240 min, and teepol (5 or 10%) for 240 min eliminates cowpea banding mosaic virus without affecting viability.[194] LMV is inactivated in polyethyleneglycol-imbibed lettuce seeds after 6 to 10 days at 40°C, or 16 days at 38°C. Dry seed treatment for 21 days at either 22, 38, or 40°C does not inactivate the virus. Treated seeds are stored following treatment.[195] Dipping urdbean seeds in 0.05% 2-thiouracil and 1% 8-azaguanine for 30 min inactivates urdbean leaf crinkle virus in seeds without affecting germination.[196]

The nematode *Aphelenchoides besseyi* can be controlled in rice seeds using 3-methyl-5-ethyl rhodanine, 3-P chlorophenyl-5-methylrhodanine, or 3-P-chlorophenyl-5-ethyl rhodanine.[197] Rice seeds soaked in 20% emulsifiable concentrate of rhodanine at 1:100 to 300 for 24 hr or 1:400 to 500 for 48 hr at 15°C can kill up to 95% of the nema-todes.[198]

5. General Guidelines for Seed Treatment
a. Use of a Good Seed-Treatment Chemical
A number of chemicals of different formulations are marketed throughout the world. A good seed treatment chemical should: (1) be effective under different agro-climatic conditions, (2) not be phytotoxic, (3) be safe to operators during handling and sowing and to wildlife, (4) not leave harmful residues in plants or in the soil, (5) be compatible with other seed-treatment chemicals, (6) be compatible with the *Rhizobium* inoculum, if necessary, and (7) be economically beneficial.[82,83]

b. Objectives of Seed Treatment
The seeds are treated for control of seedborne inoculum. However, a 100% control may not be achieved. Seeds may be treated during export or import at quarantine stations. For quarantine, however, a 100% eradication of seedborne inoculum is desir-able. However, there are few examples where complete control of seedborne inoculum has been achieved through fungicide seed treatment. Careful attention is required in determining the inoculum, location of inoculum, and amount of inoculum and host cultivar before seeds are treated with a fungicide.[82,83] For the eradication of inoculum, highly selective fungicides, mostly systemics, are useful.[82,83] Seed treatment with 0.3% tridemorph and thiabenzimidazole is effective in eradicating *Ascochyta rabiei* from chickpea seeds,[114,199] whereas 0.3% of a mixture of 30% benomyl + 30% thiram erad-icates *F. oxysporum* f. sp. *ciceri.*[200] In peas, seed treatment with benomyl (4 to 55 kg) gives complete control of *A. pisi.*[131]

c. Selective Seed Treatment
With the development of systemic fungicides and better methods of seed analysis for seed health, it is a practice to have selective seed treatments. Healthy, good-quality seeds need not be treated if there are no pathological or germination problems. Guide-lines were developed by the Illinois Crop Improvement Association, Urbana, and may be used as a rough guide to relative seed health (Table 23).[17] Seeds that meet the criteria for healthy, vigorous seeds do not require fungicide seed treatment unless they are planted under adverse conditions at a reduced seeding rate in a seed-production field. With below average to low-quality seeds, a fungicide seed treatment would increase stands and possible yields.[17]

Table 23

SOYBEAN (*Glycine max*) SEED QUALITY RATINGS

Quality text	Healthy, vigorous seeds	Below average to low-quality seeds
Warm germination (23—25°C)	85% or more	75—84%
Cold soil emergence (15°C)	70% or more	60—69%
Vigor test (23—25°C)	74% or more	55—73%
Diseased seeds	10% or less	11—20%

From Sinclair, J. B., Ed., *Compendium of Soybean Diseases,* American Phytopathological Society, St. Paul, Minn., 1982, 104. With permission.

Treatment of wheat seeds with carboxin for control of loose smut depends upon internal infection.[178,201]

Seed treatment is needed in Europe and Scandinavia to ensure emergence of wheat seeds infected by *Leptosphaeria nodorum* (*Septoria nodorum*), but is unnecessary in New Zealand where seed infection by this pathogen is unknown.[202] Klitgard and Jørgensen[203] state that wheat seed lots are treated only when laboratory tests confirm *S. nodorum* infection. Seedborne fungi of cereals in Denmark such as *Drechslera graminea, Tilletia caries, Ustilago avenae, and Urocystis occulta,* are controlled by applying organomercury fungicides to prebasic, basic, and first generation certified seeds, and nonmercurial fungicides to second generation certified seeds.[204] *D. graminea, D. sorokiniana, D. teres, F. avenaceum, F. culmorum, F. graminearum,* and *F. nivale* are important pathogens occurring in barley seeds. The effect on yield due to seed treatment is negative if the number of seedlings with discolored roots, as determined by the Doyer filter-paper method, is below 15% and positive if above 15%. Therefore, if infection is above 15%, seed treatment is recommended.[205] Seed treatment with mercury compounds is allowed in Sweden for summer cereal seeds with 15% fungal infection.[206]

d. Dosage

Seed-treatment dosage depends upon the nature of the fungicide, inoculum, location, amount of inoculum, geographic area, physical condition of seeds, etc. The average recommended dose of a fungicide for seed treatment is 2.5 g/kg seed. For soybeans, a higher (4.5 g/kg) dosage of thiram or thiram + captan is required for maintaining a plant stand under Tarai condition of Uttar Pradesh, India.[123] *Alternaria brassicicola* in *Brassica oleracea* (cabbage) seeds with up to 61.5% infection is eradicated by application of 2.5 g active ingredient per kilogram iprodione 50% WP, but higher doses are required for higher infection levels to eradicate the fungus.[107] Wheat seed treatment with carboxin at 50 to 250 g/qt significantly decreases loose smut incidence below the control. Disease control increases with increase in treatment dose. The average disease control is 45 and 90% at 50 and 250 g/qt, respectively.[178]

e. Method of Application

To obtain maximum benefits of a seed treatment, a suitable method must be used. The selection of a method depends on the fungicide, type of infection, and crop species. *P. syringae* pv. *phaseolicola* in bean (*Phaseolus vulgaris*) seeds is controlled by soaking seeds in a streptomycin solution. Pelleting seeds with streptomycin sulfate fails to control infection because of slow absorption of streptomycin from the pellet compared with rapid uptake from a solution.[207] Mistomatic® seed treaters are prefered over dry or slurry treaters because of more uniform distribution of chemicals on seeds. Leguminous seeds often are subjected to a dry seed treatment.

f. Combination of a Systemic and a Nonsystemic Fungicide

Systemic fungicides are highly selective. Thus, a nonsystemic fungicide should be mixed with a systemic fungicide to ensure a broad spectrum of control. Such a combination inhibits or controls the important seedborne pathogens and gives protection against soilborne seed decay and seedling blight pathogens. A combination of benomyl + thiram (1:1) at 0.4% is effective for the control of seedborne *Cercospora kikuchii* and seed decay and seedling blights of soybeans.[38]

g. Location and Amount of Inoculum

The location of inoculum within seeds also governs the type of treatment. For example, control of *Ascochyta pinodes* in pea seeds is obtained by treatment even with granosan at 3 kg/ton when mycelium is found only on the seed coat, but not if found in the endosperm up to 2 mm or more.[153]

Alternaria tenuissima, Aspergillus spp., *Cladosporium* sp., *F. semitectum, Lasiodiplodia theobromae, Penicillium* spp., and *Phomopsis* sp. are isolated frequently from pigeonpea seed coats and occasionally from embryo tissues. Captan and thiram move into seed coat tissues but not into the embryo. Benomyl penetrates the seed coat and embryo, thus captan and thiram are effective only against fungi in seed coat.[208] *X. campestris* pv. *campestris* is deep seated in crucifer seed tissue and thus resists chemical treatment and is controlled by hot-water treatment. Surface contaminants yield to both chemical and hot-water treatment.[209] Hot-water treatment (30 min at 50 to 52°C) may not eradicate deep-seated infection by the bacterium in crucifers.[210]

h. Resistance to Chemicals

Continuous use of a fungicide may result in the development of resistance to that chemical by the pathogen. Lack of conrol of *Leptosphaeria nodorum* (*S. nodorum*) may be due to resistance to organomercurial (phenyl mercury acetate).[211] Oat leafspot, caused by *Pyrenophora avenae*, was controlled by seed treatment with organomercury (ethyl mercury chloride) dusts in Scotland in the 1920s, but not in the 1960s because of fungal resistance.[212] Kuiper[213] described *T. caries* resistance to hexachlorobenzene in Australia and associated it with its fungicidal specificity. Mercury-tolerant strains of *P. avenae* were reported in oat seeds.[214]

i. Phytotoxicity

Seed treatment chemicals should not be phytotoxic. Streptomycin is phytotoxic to soybean seedlings in the field but is useful for control of *Bacillus subtilis* seed decay in culture plates and in vermiculite but not in soil.[215] *X. campestris* pv. *campestris* is eradicated from Brassica seeds by soaking for 1 hr in a 500 µg/m*l* solution of either aureomycin, streptomycin, or terramycin. These chemicals are phytotoxic, but if the soak is followed by a water rinse and a soak in 0.5% (w/v) sodium hypochlorite (NaOCl) for 30 min, phytotoxicity is eliminated. The effect of NaOCl is due to its oxidizing any residual antibiotic on seeds or under the seed coats.[216]

j. Combination of Seed Treatment and Foliar Spray

Seed treatment alone may not be effective in controlling seedborne pathogens. Seed treatment followed by a fungicidal spray may be more beneficial in disease control than seed treatment alone. Metalaxyl 25 WP as seed treatment at 1 g active ingredient per kilogram seed and 1 spray 40 days after planting at 1 g active ingredient per liter (750 *l*/ha) controls both systemic infection and local lesions of *Sclerospora sorghi* in sorghum. Seed treatment alone does not protect against late systemic infection of main shoots or nodal tillers, or against local lesions on leaves.[217]

k. National Laws

Use of seed-treatment chemicals is regulated by national laws. In the U.S., federal and state regulations ban the use of mercury fungicides. In the Netherlands, organo-mercury compounds are restricted to cereal seed for multiplication purposes.[4]

l. Seed Coloration

Seeds may be dyed by seedsmen to identify seed lots or improve seed appearance. The dyes are used at rates that are safe.[218]

m. Use of Nonaqueous and Aqueous Solvents for the Infusion of Fungicides in Seeds

With surface-type seed treatments, significant amounts of chemical are lost to the surrounding air and soil, thus reducing its efficiency and polluting the environment. Also, fungicides are not always uniformly distributed on seed surfaces and may not be absorbed by seeds. These problems can be overcome by incorporating chemicals into seeds using volatile solvents. The solvents carry antimicrobial materials into the seed coat and then evaporate when air dried, thus leaving the fungicide or antibiotic inside the seeds. No water is used, thus the seed coat remains firm and unswollen. Dichloro-methane (DCM) is used to introduce germination inhibitors,[219] amino acids,[220] penicil-lin,[221] erythromycin,[222] and systemic fungicides in dry seeds.[223] In addition, acetone and polyethyleneglycol (PEG) have been used to incorporate antibiotics and systemic fungicides into soybean[221,224,225] and vegetable seeds.[226] Heydecker et al.[227] reported that seeds of several vegetables or ornamentals soaked in aerated solutions of PEG partially imbibe water and freshly soaked seeds fail to germinate. PEG is a better solvent than DCM because of its low cost and nontoxicity to humans.[228]

Systemic fungicides are infused easily into soybean seeds and are effective for a relatively long time. In addition, less fungicide is used, and, because infusion reduces the amount of material on the seed coat, fungicides are less likely to interact with *Rhizobium* inoculum. Infusion of systemic fungicides with acetone in soybean controls damping-off caused by *Phytophthora megasperma* f. sp. *glycinea*. Infusion of beno-myl with DCM reduces the incidence of seedborne *Phomopsis* spp. Soaking soybean seeds in a solution of PEG 6000 containing either streptomycin sulfate or penicillin G inhibits *B. subtilis*.[17,225,229] Incorporation of quintozene with acetone or DCM controls storage decay of soybean seeds by *A. ruber*.[226] Fungicide solubility is not related to seed uptake. Benomyl, sisthane, and thiabendazole are detected in soybean seeds treated with acetone, DCM, or PEG. However, carboxin is detected only in seeds treated with PEG. Greater reduction of seedborne *Phomopsis* sp. occurred when be-nomyl and sisthane are infused with PEG than with either acetone or DCM. No treat-ment is 100% effective. Acetone, and to a greater extent, DCM kills *Phomopsis* sp. borne within the seed coat, however, they damage exposed cotyledonary tissues. The efficacy of any treatment depends on fungicide and solvent carrier combination.[230] Benomyl, fenapronil, or thiabendazole but not chlorothalonil or quintozene are carried into soybean seed coats on soaking seeds in a mixture of each fungicide with either benzene or ethanol for 30 min or longer. None of the fungicides are carried into the endosperm.[231]

The role of the solvent in infusion of soybean seeds with fungicides is complicated. The close association of fungicides, solvent, and seed allows a variety of interactions. DCM and to a lesser extent acetone kill exposed cotyledonary tissues in seeds with a cracked seed coat. These solvent carriers also appear to penetrate seeds through fungal lesions in the seed coat and subsequently increase the level of damage associated with the lesions. Seeds that have been immersed in these solvents show saprophytic growth and increased recovery of *B. subtilis*, in part because of the damage caused by the solvents. At present, the infusion of fungicides into soybeans is experimental and should not be used for commercial planting.[17]

Table 24
EFFECT OF VARIOUS SEED-TREATMENT
CHEMICALS ON SOYBEAN NODULATION

Decrease	No adverse effect	Increase
Captan[238]	Aureofungin[242]	Benomyl[239]
Carboxin[239]	Benomyl[241]	Dichlone[247]
Chloranil[240]	Captan[241,244,245]	Molybdenum[247]
Copper fungicides[241]	Carbendazim[244,246]	
1% Hg as ethyl mercury	Carboxin[238,247]	
chloride & phenyl	Chloranil[248]	
mercury acetate[242]	2.5% Hg as methoxyethyl	
Oxycarboxin[239]	mercury chloride[242]	
Quintozene[238]	Thiabendazole[249]	
Thiram[242,243]	Thiram[122,238,241,244,250]	
	Zineb[242,247]	

Solvents for the incorporation of fungicides have been used in other crops. Soaking asparagus seeds in 1.5% benomyl in acetone for 24 hr (on a shaker) followed by surface sterilization in a 10% Clorox® (NaOCl) solution controls seedling infections by *F. moniliforme* and *F. oxysporum*.[232] More fungicides (carboxin, ethazol) are found within or upon embryos in cottonseeds treated with acetone infusion than when applied directly.[233] Infusion of fungicides into pea seeds using acetone requires less fungicide than when the fungicide is applied as a slurry to control seed rot or damping-off.[234,235] Acetone, benzene, or ethanol are equally effective in carrying benomyl and thiabendazole into bean seeds. It overcomes the disadvantage of treating seeds in aqueous fungicide suspension or with powder fungicides.[236] DCM is superior to trichloromethane or carbontetrachloride in facilitating benomyl, TBZ, and thiophanate movement into bean seeds. There is interaction between seed type, solvent, and fungicide in fungicide infusion into the seed coat.[236]

n. Seed Treatment and Rhizobium Inoculum
Proper plant stand and nodulation are two important factors in successful leguminous crop production. Seeds often are treated with a fungicide and then *Rhizobium*. Studies made on the compatibility of seed-treatment fungicides and *Rhizobium* inoculum in soybeans were reviewed by Agarwal and Sinclair[237] (Table 24). Copper fungicides, oxycarboxin, and quintozene adversely affect nodulation while aureofungin (a heptaine antibiotic, *N*-methyl para-amino acetophenone and mycosamine), benzimidazoles (benomyl, carbendazim, thiabendazole), and zineb have no adverse effect. Reports on the efficacy of seed treatment with captan, carboxin, chloranil, dichlone, mercurials, and thiram are inconsistent.

6. How to Avoid Damage to Rhizobium Inoculum
The application of *Rhizobium* inoculum to seeds does not ensure nodulation, and the presence of nodules does not always mean a high level of nitrogen fixation. A number of factors such as soil pH, moisture, texture, *Rhizobium* strain, and method and time of inoculation can affect nodulation.[251] Captan, carboxin, and thiram may be compatible with *Rhizobium* in plate-count tests, but in field studies on nodule counts at 2 weeks, only thiram is found to be compatible on soybeans.[238] The adverse effect of chemicals may be in the form of reduction of total number of nodules and/ or size (weight) of nodules. The damage to nodulation by seed treatments can be reduced by:

Use of a peat-based medium — Application of inoculum in a peat base immediately prior to planting using the wet method and sucrose in the inoculating fluid reduces damage to *Rhizobium*. Quintozene, quintozene plus ethazol, and thiram are least toxic to *Rhizobium* when applied to cowpea seeds before inoculation with a peat inoculant.[252]

Maintenance of high soil moisture at planting time —

Use of higher doses of inoculum — Higher doses of inoculum may offset fungicide injury. When captan and zineb are used, increased *Rhizobium* levels do not increase nodulation. However, using carboxin, increased nodulation is observed with increased inoculum.[237]

Application of *Rhizobium* at planting time — Treated seed should be sown immediately after inoculation. Curley and Burton[238] found that 40% of *Rhizobium* applied to seeds survived as long as 1 hr. Captan and carboxin reduced viable *Rhizobium* by less than 20% in 1 hr, whereas quintozene killed 78% of the *Rhizobium*. Thiram had no adverse effect on *Rhizobium* survival. Carboxin was found compatible when seeds were planted within 4 hr of inoculation but not when held for 24 hr.[238]

Use of fungicide-resistant mutant strains of *Rhizobium* — *Rhizobium* strains differ in compatibility with fungicides. A combination of a fungicide with a tolerant strain may give nodulation higher than the control. Odeyemi[253] reported that mutant strains of *R. meliloti* or *R. phaseoli* plus thiram on alfalfa, or plus chloranil on beans, or a cowpea *Rhizobium* strain plus dichlone on cowpeas produced 100% more nodulation than wild strains. *Rhizobium* L × 717 plus captan, L × 717 plus carboxin, L × 718 plus thiram, or L × 718 plus quintozene seed treatment resulted in higher nodulation than the control (no fungicide) in soybeans.[250]

Use of granular *Rhizobium* inoculum — Granular inoculants may be used to prevent fungicide damage to *Rhizobium* inoculum since the inoculum does not come in direct contact with the fungicide. Granular applicators are used to drop inoculant near peanut seeds during planting.[253] Major drawbacks are that inoculum is expensive and planters must be fitted with a granular applicator.

Use of organic solvent as carrier for incorporating fungicide into seeds — Sinclair[229] summarized data showing that using anhydrous DCM as a carrier to introduce chemicals into seeds facilitated the movement of methylbenzimidazole carbamate (MBC) and thiabendazole into dormant soybean seeds. The solvent carried the antimicrobial material into the seed coat and evaporated when allowed to air dry, thus leaving the fungicides in the seed coat. The chances of interference of *Rhizobium* inoculum by the fungicide inside the seed is less than with dry or slurry seed treatment.

C. Mechanical Method

Seed lots may harbor inert material, colonized plant debris, pathogenic fungal propagules, etc. In addition, the seeds may be discolored, distorted, small, or enlarged due to infections. Such seed lots should be cleaned mechanically. The use of clean seeds can reduce seedborne inoculum but does not eliminate it. The use of clean seeds for control of cucumber mosaic and TMV was stressed by Bewley and Corbett.[254]

Processing, screening, and sieving can remove all inert material, plant debris, and fungal fruiting bodies. *F. oxysporum* f. sp. *betae* is an external contaminant in sugar beet seeds. Commercial processing of seeds reduces the amount of infested seed and seed transmission.[255] LMV occurs in seeds of light weight, and such seeds can be separated from heavier, noninfected seeds.[256] Separation of LMV-infected seeds can be done using a vertical air stream.[257] The removal of small barley seed reduces BSMV.[258] The removal of pea seed with growth cracks reduces pea seedborne mosaic virus in a seed lot from 33 to 4%.[259]

Removal of the infested pulp from vegetable seeds at extraction helps to reduce seedborne inoculum. *Corynebacterium michiganense* pv. *michiganense* is eradicated from tomato seeds by extracting pulp with hydrochloric acid (HCl) followed by drying in a tumbler drier for 3 hr at 66°C.[260] When extracted with HCl (25 ml commercial HCl added to pulp from 2.27 kg fruit), seeds of the tomato cultivars Potentate and V 548 did not show seed transmission of TMV.[261] Cleaning seeds removes fruit pulp remnants that carry TMV in tomato and cucumber green mottle virus in cucumber seeds.

Seeds with the wheat gall nematode, *Anguina tritici*, have a lower specific gravity and can be separated from noninfected seeds by dipping the seed in a 20% NaCl solution. The seeds are dried and treated with a suitable seed protectant. Sunflower seeds are stirred in warm water (2 *l*/kg seed) for 15 min, then floating seeds are removed and dried. No sclerotia of *Sclerotinia sclerotiorum* are found with such seeds.[262] Soybeans readily absorb water, which makes the seed coat expand, become slippery, and pull away from the embryo. A nonaqueous mixture of glycerine and polyethyleneglycol 400 (60:40) can be used to separate lightweight, infected from the heavier, uninfected soybean seeds.[263] The separation of light, poor-quality discolored cottonseeds by flotation in water after delinting in sulfuric acid helps in removing seeds infected with *X. campestris* pv. *malvacearum*.[264]

D. Physical Methods

Physical methods or thermotherapy are some of the oldest methods used for control of seedborne pathogens. Thermotherapy is used for control of seedborne pathogens when seed-treatment chemicals are not available. Jensen demonstrated in 1882 that the hyphae and spores of *Phytophthora infestans* in infected potato tubers are killed in a 4-hr hot-air treatment at 40°C. In 1888, he found that hot-water treatment of dry seeds controls *Ustilago hordei* and water soaking controls *U. hordei* and *U. tritici*.[204] This method was followed in Denmark and later adopted in the U.S. on the recommendation of Swingle in 1892.[83]

The principle of thermotherapy is that microorganisms are killed or viruses destroyed at temperatures not injurious to seeds.[265] Seeds with a low moisture content are ideal for heat therapy. Seeds with a high moisture content are killed at lower temperatures than those with a low moisture content. Various theories have been proposed on the cause of seed death from exposure to high temperatures:[265] denaturation of proteins, lipid liberation, hormone destruction, tissue asphyxiation, depletion of food reserves, and metabolic injury with or without accumulation of toxic intermediates. The methods used for the application of heat to seeds are as follows:

1. Hot-Water Treatment

The use of hot water is widely used to control seedborne pathogens, especially bacteria and viruses, and includes the following steps:[265]

Selection of seeds — Seeds which can withstand hot-water treatment are preferred. Legumes and soybean seeds cannot be subjected to hot-water treatment because their seed coat swells easily and sloughs off. Pathogens easily controlled by fungicide seed treatment usually are not subjected to hot-water treatment. Hot-water treatment is recommended for seeds with deep-seated infections which cannot be eradicated by other means.

Presoaking the seed — Seeds may or may not be soaked in water before treatment depending upon the pathogen and crop. Presoaking is done to replace air between embryos and seed coats with water which is a better heat conductor. The water soak may stimulate pathogen growth which becomes more heat susceptible. The duration of soak (4 to 12 hr) may not be sufficient for water penetration, hydration of tissue, and initiation of pathogen growth in all seeds.

Preheating — After soaking in cool water, seeds are heated for 1 to 2 min at 9 to 10°C below the temperature of the final treatment. This is done to counter the cooling effect of soaking in cool water.

Hot-water soak — The temperature and time required varies with the pathogen and crop. Timing must be precise, otherwise seed viability is lost. A large volume of water helps maintain a constant temperature. Seeds should be packed loosely in porous bags, screen boxes, or frames with sufficient provision for an ample water flow. Water in tanks may be heated by thermostatically controlled, electric immersion heaters or by steam pipes.

Cooling — Treated seeds are spread out for cooling and drying immediately after heat treatment.

Drying — Seeds are dried quickly to prevent sprouting.

Post-treatment — The application of a seed protectant fungicide maintains seed viability and avoids seed decay due to soilborne microflora.

Hot-water treatment is not applicable for seeds which absorb water resulting in seed coat rupture, such as legume seeds, or exude mucilaginous materials which stick seeds together on drying, such as flax seeds. Carbon tetrachloride and oils are suggested as alternatives to hot water treatment. Watson et al.[266] tested various nonaqueous fluids to find a medium for heating seeds of large-seeded legumes. They found green and lima bean seeds survived longer when treated in motor oil at 90°C than in water at 90°C, and seeds heated for 60 min in boiling CCl_4 (76.8°C) also survived. Soybean oil is used for heating soybean (*Glycine max*) seeds to control seedborne fungi, especially *Phomopsis* spp.[267] Treatments that decreased *Phomopsis* and increased germinable pathogen-free seeds ranged from 5 min at 70°C to 10 sec at 140°C. Seeds placed in soybean oil at 21°C do not imbibe oil, swell, or slough their seed coats.[267]

Cabbage (*Brassica oleracea*) seed thermotherapy often results in injury expressed as reduced or delayed germination. PEG treatment of heat-damaged cabbage seeds increased germination and emergence and restored germination rate and emergence. Cabbage seeds are dipped in PEG 6000 (305 g/ℓ water) for 14 days at 15°C. The treatment stimulates a physiological repair mechanism which may operate in fully imbibed but dormant seed. There is a possibility that the pathogen also may recover, but pathogens do not have the same metabolic reserves.[222]

2. Hot Air Treatment

Hot air is less effective than hot water in the control of seedborne pathogens. The advantage of hot-air treatment is that it is easy and seeds are less damaged.

3. Solar Heat Treatment

Solar heat treatment is effective for controlling loose smut of wheat using small quantities of seeds during hot summer months. During the last week of May or first week of June in north India, where summers are hot (above 35°C), seeds are soaked in water for 4 hr in the morning and then dried in the sun before storing.[268]

4. Aerated Steam Treatment

The use of aerated steam is safer than hot water and more effective than hot air. The heat capacity of water vapor is about half that of water and 2.5 times that of air; hence, the temperature and time required may be higher than that of hot water and lower than that of hot air.[265] Drying of seeds is easier, loss in germination is low, temperature control is easy, seed coats of legumes remain intact, and flax seed do not become sticky. *Itersonilia pastinacea* in parsnip plant debris associated with seeds can be inactivated with a steam/air mixture for 30 min at 45.5°C.[269]

5. Radiation

Electromagnetic radiation has been studied for control of seedborne pathogens. Soybean seeds were used to test the efficiency of disinfection by radiation, high voltage of electric currents, ultrasonic radiation, and very high frequency (VHF) radio waves. On passing electricity at 4 kw/g for 30 sec through seeds, bacterial infection decreased from 5.9 to 2%, the degree of disinfection depending on voltage and exposure. Bacterial infection in plants grown from seeds when exposed to ultrasonic radiation decreased with exposure of 21.3 kc/sec for 15 min. Cotyledon bacteriosis (*Pseudomonas syringae* pv. *glycinea*, and *P. solanacearum*) and angular leafspot (*P. syringae* pv. *glycinea*) were 16.5 and 40.4%, respectively, compared with 30.8 and 73.2% in control plants. No specific results were obtained with VHF waves, though angular leafspot bacteria were inhibited. Germination of seeds was not affected.[270] Treating tobacco seeds with 625-W microwave radiation for 20 min eliminates *Erwinia carotovora* pv. *carotovora* without affecting germination. The number of infected seeds declined by 68 and 99% by a 10- and 15-min treatment, respectively. Infected seeds had a 5.3% moisture content.[271]

Gamma irradiation of barley seeds from a 1800−C ^{60}Co source from 0 Roentgens to a lethal dose increased the ratio of healthy plants over plants infected with barley stripe mosaic. This may be due to a lethal effect on infected seeds, on the virus in some seeds, or on symptom expression in seedlings.[272] Radiation of *Prunus* seeds at higher than 20,000 rad caused reduction in seed transmission of necrotic ringspot and prune dwarf viruses; seed viability also was reduced.[273]

6. Factors Which Govern Heat Therapy

The factors influencing results of heat therapy are seed moisture content, dormancy, age, vigor, and physical condition; as well as cultivar susceptibility, and location and amount of inoculum. Reduction of seedborne inoculum without affecting seed viability is not always attained.[265]

7. Application

Hot-water treatment used correctly can eliminate most seedborne bacteria and fungi without affecting germination. It is the only reliable alternative method to the use of mercurial seed treatment which is banned in most countries. Hot-water treatment can be used on seeds of cabbage, carrot, cucumber, eggplant, lettuce, pepper, radish, spinach, tomato, and turnip. The procedure involves placing seeds in a loosely woven cotton bag to half full, warming them in water for 10 min at 38°C before placing the bags in a waterbath at the recommended time and temperature required. Seeds are dipped in cold water immediately after treatment and dried. The seeds are treated with seed protectant before planting. Diseases which are controlled with hot water are given in Table 25.

Some viruses within seeds can be inactivated by thermotherapy, but most viruses in dry seeds are resistant to heat. Reddick and Stewart[285] found that bean seeds treated with dry heat for 10 min at 80°C or up to 40 min at 75°C had reduced survival of BCMV, but still had 50% transmission. Similarly, hot-water treatment for 15 min at 70°C showed transmission of BCMV. In every case the virus survived where seeds survived. TRSV in soybean seeds is not inactivated by heat treatment without destroying seed viability.[286] By treating dry lettuce seeds above 100°C, LMV could be reduced, but germination is reduced so much that the effort is not worthwhile.[287] There is no loss of BSMV in barley seeds treated for 30 min at 130°C.[288]

Exposure of seeds to high temperature for short time periods is not effective in eliminating viruses from seeds. Longer periods at lower temperatures are effective in

Table 25

CONTROL OF IMPORTANT SEEDBORNE PATHOGENS THROUGH HOT-WATER TREATMENT OF SEEDS

Crop	Disease	Pathogen	Treatment*	Ref.
Arachis hypogaea (groundnut, peanut)	Testa nematode	*Aphelenchoides arachidis*	15 min cool water 60°C for 5 min	274
Brassica spp. (broccoli, cauliflower, Chinese cabbage, collard, kole, kohlrabi, rape, rutabaga, turnip), *Cucumis sativus* (cucumber), *Daucus carota* (carrot)	Black rot	*Xanthomonas campestris*	50°C for 30 min	109
Brassica spp., (brussels sprouts, cabbage), *Solanum melongena* (brinjal, eggplant), *Lycopersicon esculentum* (tomato), *Spinacia oleracea* (spinach)	General seed protectant General seed protectant	pv. *campestris* General seed protectant	50°C for 20 min 50°C for 25 min	275 275
Brassica spp. (mustard, cress, radish)	General seed protectant		50°C for 15 min	275
C. sativis (cucumber)	Seedling blight	*Pseudomonas syringae* pv. *lachrymans*	50°C and 75% R.H. for 3 days	276
Cyamopsis tetragonoloba (cluster bean, guar)	Blight	*X. campestris* pv. *cyamopsidis*	56°C for 10 min	277
Dipsacus spp. (teasel)	Stem eelworm	*Ditylenchus dipsaci*	1 hr at 50°C or 2 hr at 48.8°C	278
Latuca sativa (lettuce)	Leafspot	*X. campestris* pv. *vitians*	70°C for 1 to 4 days hot air	279
Lycopersicon esculentum (tomato)	Black speck	*P. syringae* pv. *tomato*	52°C for 60 min	140
Nicotiana tabacum (tomato)	Hollow stalk	*Erwinia carotovora* pv. *carotovora*	50°C for 12 min	280
Oryza sativa (paddy, rice)	Udbatta White tip	*Ephelis oryzae* *A. besseyi*	54°C for 10 min 24 hr cool water 51—53°C for 15 min	281 282
Pennisetum typhoides (pearlmillet)	Downy mildew	*Sclerospora graminicola*	55°C for 10 min	283
Tropaeolum majus (nasturtium)	Fascians disease	*Corynebacterium fascians*	1 hr cool water, 51.7°C for 30 min	284

*   R. H. = relative humidity.

reducing certain viruses. TMV is eliminated from tomato seeds after 2 days at 50 to 52°C followed by 1 day at 78 to 80°C.[289] Treatment of tomato seeds for 22 days at 72°C reduces seed transmission of TMV without loss in germination.[290] Heating dry tomato seeds for 3 days at 70°C also reduces TMV.[291] In East Germany, dry-heat treatment of tomato seeds for 24 hr at 80°C is officially advocated for control of TMV.[292,293] Cucumber green mottle virus can be removed from infected cucumber seeds by hot-water treatment for 3 days at 76°C.[294] In a commercial trial, 4500 symptom-free cucumber plants were raised from previously infected seeds treated for 3 days at 70°C.[295] Cowpea banding mosaic virus can be reduced in cowpea seeds by hot-water treatment for 40 min at 45°C, or for 20 min at 50°C; with hot air for 50 min at 55°C, or for 20 min at 65°C; or a dry treatment for 15 min at 65°C followed by 2, 4, or 8 days at 30°C.[194] Cowpea mosaic virus is inactivated in cowpea seeds when freshly infected seeds are exposed for 4 days to 30°C or for 15 min at 55°C followed by 4 days at 25°C.[296] Hot-air treatment of vegetable marrow seeds for 2 days at 70°C or for 4 weeks at 40°C or hot-water treatment for 6 min at 55°C eliminates seedborne cucumber mosaic virus.[297]

Soaking *A. tritici* galls for 10 min at 55°C kills 97% of the larvae, whereas 83% of the larvae are killed in dry sand after 10 min at 55± 1°C. After 15 min, all larvae are dead in both treatments.[298]

8. Limitations of Thermotherapy

Thermotherapy for the control of seedborne pathogens has limitations:[210,299,300] (1) there are chances of reduced germination, (2) seed coats may swell and split as in legumes, (3) seeds may stick together due to the release of mucilagenous substances, (4) it is difficult to maintain the temperature and time requirements, (5) it may not be as effective as a fungicidal seed treatment, (6) it is difficult to process, (7) the mode of action is not definitely known, and (8) deep-seated infections may not be eliminated completely.

V. CERTIFICATION

Certification ensures that seed lots meet certain quality standards and that the history of each lot is traceable. It involves testing seeds before sowing and after harvest; and crop inspection for compliance with standards, including isolation and freedom from weed seeds and diseases.[4] Certification programs follow breeders' seed through successive multiplication. Through certification, certain seedborne pathogens have been controlled, and spread to new areas checked. Presently, there is more emphasis on seed testing for seedborne pathogens in seed-certification programs than on disease incidence in the field. Field inspection is useful in rejecting seed lots with a high incidence of seed-transmitted pathogens, but apparent absence of disease does not guarantee pathogen-free seeds. The causal agents of anthracnose, bacterial blight, frogeye leaf spot, Phomopsis seed decay, pod and stem blight, stem canker, and soybean mosaic may be present in soybean without producing symptoms (asymptomatic) until plants begin to mature.[17] Cafati and Saettler[301] emphasized that tests to detect seedborne *Xanthomonas campestris* pv. *phaseoli* should be included in any production program for certified blight-free *Phaseolus acutifolius* and *P. vulgaris* seeds because seed transmission of the bacteria occurs in symptomless seeds from symptomless pods of both resistant and susceptible genotypes.

Loose smut of barley and wheat is controlled through seed-certification programs in India, Scotland, and Sweden.[1,178,302] Certification is based on seed production in isolation, field inspections, laboratory evaluation of seedborne infection by the embryo-count method, and seed treatment with a systemic fungicide. In England, the standard for field approval from infection by wheat and barley is 1 smutted ear per 10,000 ears. Lower standards are applicable to susceptible cultivars of both barley and wheat, but not more than 1 in 2000 in commercial crops and 1 in 5000 in crops intended for multiplication.[303] Field inspection for loose smut was discontinued in many places in favor of a laboratory embryo-count test, which permits 0.2% infection.[304] The embryo test for detecting loose smut hyphae in barley seeds is used at the Official Seed Testing Station for Scotland, East Craigs. The test is carried out on samples of commercial and on farm-saved seeds submitted for certification. A standard of 0.2% infection, determined by a laboratory test on 1000 embryos, is set in conjunction with the Scottish Central Seed Certification Scheme.[302]

The embryo test for detection of loose smut in wheat seeds also is used by the Uttar Pradesh Seeds and Tarai Development Corporation Ltd., India. Seed lots with up to 0.5% infection are certified without seed treatment and those between 0.51 to 2% are treated. Seed lots with more than 2% loose smut infection are not used for multiplication. Use of the embryo test and the necessity of treating seed with more than 0.5% infection has restricted loose smut infection.[1,178] A standard of no *Plenodomus lingam* in 1000 prebasic and basic seeds for multiplication has been proposed for Brassica seed certification in the U.K. Similar standards were set for *Phoma betae* on red beet, *Septoria apiicola* and *P. apiicola* on celery, *Colletotrichum lindemuthianum* and *Pseudomonas syringae* pv. *phaseolicola* on bean, and *Ascochyta fabae* on broadbean.[4] Infection of *A. fabae* in British-grown commercial bean seed is reduced greatly by seed selection.[305]

A rigid seed-certification program contributed to elimination of seedborne bacteria from bean seed stocks in Idaho. Using serological or greenhouse evaluations or both, seed samples can be shown to be free of seedborne bacteria.[306] The Michigan bean seed program involves production of early multiplications in California and Idaho with a standard, permitting fields with 0.005% blighted plants (*P. syringae* pv. *phaseolicola*) and zero tolerance in seeds using a seedling infection test with leachates from 2.27 kg seeds.[307] In the Canadian Scheme, disease standards for fuscous blight (*X. campestris* pv. *phaseoli* var. *fuscans*) in beans include zero tolerance in a test on a 5-kg sample. If blight is detected, the resistant foundation seed must show a negative result in a test on a sample of 3.4 kg.[4,308]

Control of seedborne viruses such as LMV in lettuce and BSMV in barley is achieved through certification programs. Virus-free planting material is obtained from a disease-free crop. An inspection for virus infection is carried out regularly and infected plants are rogued. The selection of healthy plants based on absence of symptoms may be misleading in cases when viruses are masked. In such cases, the presence or absence of a virus is confirmed using other tests.

The use of LMV-free seed stocks of lettuce has controlled LMV in the Imperial Valley, California since 1969. The program reduced yield losses by 95 to 100%. Lettuce seeds are grown in insect-free greenhouses, infected plants are rogued, and seeds are produced from virus-free plants. Seeds are certified by growing on or infectivity tests. Seed lots with no infection in 30,000 seeds are used for planting. In Europe, tolerance for this virus is 0.1%.[309-312]

The incidence and severity of BSMV in barley have been reduced through seed certification and planting virus-free seeds in the U.S. since 1970. Seeds are tested for BSMV in barley embryos using serological methods.[313] Two certification programs exist. A complete generation certification program initiated in Montana in 1966 involves the production of foundation, registered, and certified seeds. Seeds are harvested from inspected fields with no diseased plants, and samples are tested for the virus using the sodium dodecyl sulfate (SDS)-disk test. In North Dakota, a limited generation certification program is followed. Foundation seed fields are examined and seed lots are tested for BSMV using the latex flocculation method. Both programs have prevented millions of dollars of crop loss due to BSMV.[314] In Canada, pea seedborne mosaic virus-infected pea lines are screened and eradicated on the basis of routine examination of seed lots.[315] Seed production plot inspection also is used in certifying for absence of pea-early browning virus in peas in the Netherlands. Over 1 diseased plant per 100 m² renders the seed unfit.[316] Soybean plants are inspected five times for SMV, and screened using ELISA. Infected plants are rogued. Up to 20 leaflets per line, 1 from each of 20 plants, are tested. Lines giving a positive ELISA reaction are reinspected, rogued, and assayed again. Seeds with negative ELISA are used for further multiplication.[317]

A. Setting Certification Standards

Most certification standards are based on field inspections and/or seed health testing. Problems with standards based only on field inspections are that: (1) evaluation of field infection depends upon the skill of certification personnel, (2) the crop may be infected but the seeds may not be, (3) symptoms may be masked, making it difficult to detect the disease in the field, and (4) the correlation between disease incidence and seed infection is not known for a majority of seedborne pathogens. Another problem is certification standards based on percent of seedborne infection. Data often are lacking about the role of seedborne inoculum in disease development and subsequent yield losses (i.e., economic thresholds) and this is a major reason for lack of seed-certification standards for seedborne pathogens.

Setting certification standards on seed health testing is the most appropriate approach for a majority of plant pathogens because it is the level of seed infection which govern the transmission of pathogens through seeds. The permissible limit may vary depending on the following factors.[318]

1. Relative Role of Seedborne Infection in Disease Development

Certification standards vary depending on their potential rate of establishment, infection, and spread. For the majority of seedborne pathogens, information only on their seedborne nature is available. Transmission rate and spread in the field are not well understood. Under favorable conditions, 12 bean seeds per acre infected with *P. syringae* pv. *phaseolicola* can cause severe epiphytotics of bacterial blight.[319] Only 0.1% seed transmission of LMV leads to severe field infection when aphid vectors are active.[320] Therefore a tolerance of 0.003% is allowed in California for this virus.[321] A close correlation exists between *Ustilago tritici* infection of seeds and loose smut development in barley and wheat.[178,302,322] Thus, disease development can be measured by seed infection. Seed transmission becomes important if there is potential for spread of a pathogen from infected seeds. The tolerance must be near zero in such cases.

2. Perpetuation of the Seedborne Inoculum by Other Means

Many seedborne fungal and bacterial pathogens are soilborne. If a pathogen is only seedborne, strict certification standards can be followed and infected seeds should not be sent into areas where the pathogen has not been detected. But, if a pathogen is soilborne, its seedborne nature becomes secondary in disease development. In such cases certification standards may be relaxed.

3. Factors Affecting Seed Transmission

The environment plays an important role in seed transmission. This has been observed in karnal bunt of wheat (*Neovossia indica*) in the Nainital Tarai of India. During the 1969—70 harvest, seedborne infection varied from 0 to 7.5% in different cultivars, while in 1970—72 its incidence was less than 1%, and in 1974—75 the incidence reached 50%.[323,324] In Canada, depending upon the rate of seed infection, the tolerance limit for *Ascochyta* spp. on peas varies from 2 to 6%.[321] Therefore, seed-certification standards may vary from year to year for certain pathogen depending upon seed infection.

4. Economic Loss Due to Seedborne Pathogens

Losses caused by seedborne pathogens vary depending upon virulence, host range, and environment. If a pathogen causes heavy losses, the level of infection allowed for certification should be low. The role of seed infection, ratio of subsequent transmission in the field, and yield loss should determine seed-certification standards for different pathogens.

5. Planting Area

Certification standards also depend upon the planting area. Some pathogens may not infect a crop in certain areas due to unfavorable environmental conditions. For such areas, a relaxation in seed-certification standards can be allowed. However, in areas where infected seeds may serve as a source of soil contamination, or where the infection can occur, only pathogen-free seeds should be planted.

6. Influence of Seed Treatment on Seedborne Infection

If a seedborne pathogen can be eliminated by seed treatment, then certification standards should be relaxed. Most chemical seed treatments are not effective against bacterial and viral pathogens and certification standards are necessary for them.

7. Seed Processing Procedures

A number of plant pathogen propagules can be separated from healthy seeds during processing, such as seeds (*Cuscuta, Orobanche*), sclerotia (ergot, *Sclerotium, Sclerotinia*, etc.), and galls (ear cockle), etc. For such pathogens, seed processing must be taken into consideration while developing certification standards.

VI. PLANT QUARANTINE

Quarantine is derived from the Latin word *quarantum*, meaning 40. It refers to the 40-day period of detention of ships arriving from countries with bubonic plague and cholera in the Middle Ages. The first such quarantine was imposed in Venice in 1374.[325] Present quarantine laws now include plants. Plant quarantines, promulgated by a government or group of governments, restrict entry of plants, plant products, soil, cultures of living organisms, packing materials, and commodities, as well as their containers and means of conveyance to protect agriculture and the environment from avoidable damage by hazardous organisms. They exclude dangerous organisms while permitting plants and plant products to enter.[326] The term "exclusion" conveys this objective more clearly than "plant quarantine". Exclusion relates to keeping organisms out; plant quarantine relates to keeping plants out. This concept has been recognized by the California Department of Agriculture, which employs a "Detection and Exclusion Officer" rather than a "Plant Quarantine Officer".[326]

The importance of plant quarantines has increased because of the increase in exchange of seeds or grains for consumption along with better means of transportation. The international exchange of plants or their parts is practiced widely to improve the crops of a country and their genetic base.[327] In addition, shiploads of grains for consumption or large quantity of seeds for direct sowing is practiced in many countries.[328] Even minute quantities of soil and plant debris contaminating true seeds can disseminate pathogens.

A large number of plant pathogens have spread over the world through seeds.[304] Wheat bunt (*Tilletia caries*) appeared for the first time in the Sacramento Valley, California in 1854 on plants grown from seeds imported from Australia.[328] Peanut rust (*Puccinia arachidis*) was introduced into Brunei on peanuts imported for consumption but used for planting.[329] *Marasmius perniciosus* (witches broom of cacao) was introduced in 1974 and 1975 into South America with seeds from Trinidad. The seed produced up to 70% infected seedlings.[330] Peanut rust was introduced from Brazil to the U.S. on peanut seeds.[331] In 1942, bacterial canker of tomato (*Corynebacterium michiganense* pv. *michiganense*) was introduced into England with seeds from the U.S.; *Gloeotinia temulenta* on ryegrass seeds from New Zealand to Oregon, in 1940; and *Xanthomonas campestris* pv. *campestris* on cabbage seeds from Europe to India.[332] Two rice pathogens, *X. campestris* pv. *oryzae* and *X. campestris* pv. *oryzicola*, once confined to Asia, are established to west Africa and Brazil where none of the local cultivars are resistant.[333]

The race of the nematode *Ditylenchus dipsaci*, present in Sweden, does not attack alfalfa, but a new race which attacked alfalfa was introduced on imported seeds.[334]

A. National and International Regulations

The first plant-quarantine law was passed in 1873 in Germany to prohibit importation of plants and plant products from the U.S. to prevent the introduction of the Colorado potato beetle, and in 1877, the United Kingdom Destructive Pests Act prevented the introduction and spread of this beetle. In 1891, the first plant-quarantine measure was initiated in the U.S. by setting up a seaport inspection station at San Pedro, Calif., and the first U.S. quarantine law was passed in 1912. The Federal Plant Quarantine Service was established in Australia in 1909.[325] In India, a Destructive Insects and Pests Act was passed in 1914. Since then, most countries have formulated quarantine regulations.

On a global basis, the first International Plant Protection Convention (the Phylloxera Convention) was signed, in 1881, with the objective of preventing the spread of severe pests. This convention was amended in 1889, 1929, and 1951. The International Plant Protection Convention (IPPC or Rome Convention) under the Food and Agriculture Organization was established to prevent the introduction and spread of diseases and pests through legislation and organizations across international boundaries.[335] This convention provided a model phytosanitary certificate (Rome certificate) to be adopted by member countries. Within this convention, ten regional plant-protection organizations have been established on the basis of bio-geographical areas:[325,328,335,336] European and Mediterranean Plant Protection organization (EPPO), Inter-African Phytosanitary Council (IAPSC), Organismo International Regional de-Sanidad Agropecuria (OIRSA), Plant Protection Committee for the South-East Asia and Pacific Region (SEAPPC), Near East Plant Protection Commission (NEPPC), Comite Interameicano de Protection Agricola (CIPA), Caribbean Plant Protection Commission (CPPC), North American Plant Protection Organization (NAPPO), Organismo Bolivariano de Sanidad Agropecuria (OBSA), and ASEAN region grouping of Indonesia, Malaysia, Philippines, Thailand, and Singapore. The regional organizations are concerned with the coordination of legislation and regulations within their area, agreement on the quarantine objects, inspection procedures, etc. The EPPO, IAPSC, and ASEAN have taken up in detail the question of seed quarantines.[328,336]

1. Plant Quarantine in the U.S.

The control of plant introductions into the U.S. began in 1829, when the U.S. Congress allotted $1000 for the import of rare plants and seeds.[337] The office of the U.S. Patent Commissioner was authorized for the introduction of germplasm between 1836

Table 26

IMPORT OF SOME MAJOR CROP SEEDS
PROHIBITED IN THE U.S. DUE TO INFECTIONS[327]

Crop	Country	Disease
Gossypium sp. (cotton)	All	Viruses, various diseases
Sorghum vulgare (milo, sorghum)	Africa, Asia, Brazil	Smuts
Lens spp. (lentil)	South America	Rust
Oryza sativa (paddy, rice)	All	Smuts, viruses, various diseases
Triticum aestivum (wheat)	Asia, Australia, Eastern Europe	Flag smut
Zea mays (corn, maize)	Africa, Asia	Downy mildews

to 1862. With the establishment of the U.S. Department of Agriculture (USDA) in 1862, a Commissioner of Agriculture was made responsible for collection, testing, and distribution of potentially valuable plant germplasm. A section, Seed and Plant introduction, was established in 1898. This system has continued with minor changes (Figure 2).[327] Over 465,000 plants or seeds have been introduced since 1898. At present, about 7500 new introductions are made each year.[327] The Plant Protection and Quarantine program (PPQ) is planned and executed by the Animal and Plant Health Inspection Service (APHIS) of the USDA.[338] At present, three quarantine acts are in operation in the U.S.[327]

a. Plant Quarantine Act of 1912

The first U.S. federal plant quarantine law known as the Plant Quarantine Act of 1912 was passed after the establishment of white pine blister rust and chestnut blight fungi, and the citrus canker bacterium. The Act controls the introduction of exotic pests and the spread of plant parts new to U.S. and within the U.S. as a domestic quarantine.[338] In spite of the Plant Quarantine Act, pathogens such as the potato wart, wheat flag smut, and Dutch elm disease fungi, have been introduced into the U.S.[327]

b. Organic Act of 1944

This act is mainly for pest management strategies, but gives an authority for issuance of phytosanitary certificates in accordance with the requirements of importing states and foreign countries.

C. Federal Plant Pest Act of 1957

This act authorizes emergency actions to prevent the introduction or interstate movement of plant pests not covered under the Act of 1912 (Table 26).

Nearly all imported germplasm falls into one of three categories: restricted, postentry, or prohibited. Restricted germplasm is inspected and chemically treated, and can be imported easily. In the postentry category, seeds or other material, after inspection and treatment, are grown out under close observation. If no pest is found, then germplasm is released. Prohibited materials must meet certain specific requirements before being imported since they may pose a serious threat to agriculture.[327]

Postentry surveillance for the detection and interception of seedborne pathogens on introduced plants and the production of disease-free seeds are accomplished at regional plant introduction stations. These stations are operated by cooperative agreement between the USDA and land-grant colleges and universities. Here the plants are subjected to inspection, detection, postentry surveillance, and release of seeds.[339]

2. Plant Quarantine in the U.K.

The plant health legislation in the U.K. was passed as the Destructive Insects Act of 1877 primarily to prevent the entry and establishment of Colorado potato beetle. It was extended by the Destructive Insects and Pests Act of 1907 to check the entry of American gooseberry mildew (*Sphaerotheca morsuvae*) and all insects, fungi, or other destructive pests of plants. To cover bacteria and viruses as well as invertebrate pests, the act was extended as the Destructive Insects and Pests Act of 1927. These three acts were consolidated and formulated in a Plant Health Act of 1967. This act was amended by the European Committees Act of 1972.[340] The most familiar activity is the inspection of plants or produce either before export or after import.[341]

3. Plant Quarantine in India

The plant-quarantine activities in India are reviewed by Wadhi.[342] The earliest activities concerned with the introduction of plant pests and diseases with plant material was in the early 1900s. Fumigation of all imported cotton bales was required to prevent introduction of the Mexican boll weevil (*Anthonomus grandis*). Later it was realized that a large number of other pests and diseases could be introduced.

In 1914, the Government of India passed the Destructive Insects and Pests Act. Under this act, the import of plant material is channeled and incoming passengers, baggage, and/or cargo are examined. Plant material is imported for commerce and consumption, or as germplasm. The imports for commerce and consumption are examined at plant quarantine and fumigation stations within the country; eight at seaports, seven at airports, and seven at land frontiers. Research material is examined by three agencies, the National Bureau of Plant Genetic Resources, New Delhi, for agricultural and horticultural crops; the Forest Research Institute, Dehra Dun, for forest plants; or the Botanical Survey of India, Calcutta, for all other plants of economic general interest.

4. Plant Quarantine in Kenya

At Muguga, Kenya, the introduction of unwanted pathogens is prevented by growing the incoming material in isolated greenhouses; seed health testing; importing from selected countries that provide minimum risk; prohibiting import of certain crop species from certain countries; recovering healthy planting material by seed treatment, tissue culture, heat treatment, or tissue culture and heat treatment combined; and releasing only second-generation seed.[343]

B. Basic Principle of Plant Quarantine

The basic principle of plant quarantine is to check the entry and spread of potentially dangerous plant pathogens and insects imported along with the germplasm. In spite of quarantine regulations, plant pathogens have been introduced in different countries. Plant quarantine regulations have certain prerequisites. They must be[325,327,335,344]

1. Based on sound biological grounds. Only pests which pose a threat to major crops or forests should be taken into consideration.
2. Formulated to control or prevent the entry of pests and not to hinder trade or attainment of other objectives. Quarantine measures are for crop and not trade protection.
3. Derived from adequate legislation and operated solely under the law.
4. Modified as conditions change or further facts become available.

Those responsible for quarantine measures should be properly trained and experienced; and professional workers and the public must cooperate on an international scale for effective operation of quarantine regulations.

Quarantines are only one facet of domestic pest management programs, and careful integration of measures is needed to achieve maximum efficiency.

C. Problems in Plant Quarantines

Quarantines serve as a filter against the introduction of dangerous pathogens, but pathogens are still introduced. Possible reasons are that: (1) it is difficult to detect all types of infectious pathogens by conventional methods, (2) the methods may not be sensitive enough to reveal traces of infection, (3) the latent infections may pass undetected under postentry quarantine, (4) destruction of all infected or suspected material, and (5) lack of sensitive methods for testing fungicide-treated seeds.[339] Plant pathogens may be introduced on inert material such as packing material, dried root bits, plant debris, soil clods, etc. Cysts of *Heterodera schachtii* and *H. goettingiana* nematodes were intercepted on such materials.[345]

D. Organisms of Quarantine Significance

Organisms of quarantine significance may include any pathogen or pest that a government (or intergovernment organization) considers to pose a threat to the agriculture and environment of the country or region. Such organisms usually are exotic to that country or region but may include exotic strains or races of domestic strains.[326]

E. Plant Quarantine Measures

The aim of plant quarantine is to prevent the introduction of dangerous diseases and pests or a new races of a pathogen and their spread within the country. Measures suggested for effective plant quarantine are as follows.[325,328]

1. Import Control — Regulations of the Importing Country
a. Embargoes

This is the most effective measure to exclude infected plant material. However, in practice it is difficult to achieve because of more and more exchange of diverse genetic material among countries.

b. Inspection of Seed Lots

The examination of seed samples must be by the most sensitive and reliable methods for detection of dangerous pathogens listed under quarantine regulations. A sample may be subjected to more than one method. Detection of seedborne pathogens may be difficult if seeds are treated. It is difficult to eradicate infections by conventional seed-treatment fungicides. Systemic fungicides help eradicate certain seedborne pathogens, if the seeds have been treated accurately. Broad-spectrum fungicides which can eradicate diverse groups of pathogens are not available.

c. Postentry Quarantine

Because it is difficult to detect all types of seedborne pathogens by simple tests, it may be necessary to subject seeds to postentry quarantine. Seeds are subjected to a period of growth at a quarantine station under strict supervision in the importing country.[346,347] The plants are kept under close observations in semi-isolation so that any disease which appears can be detected immediately. Plants are grown under optimum conditions so that symptoms are not masked.[348,349] Pathogen-free seed then is produced from the imported seed material for distribution. Thus, valuable germplasm of introduced plants can be saved for breeding and crop improvement without any danger of introducing prohibited plant pathogens.[350,351]

d. Seed Treatment

Seeds may be treated with a suitable chemical before release for further multiplications or utilization in a breeding program as an additional safety measure against chance introduction of a pathogen.

2. Export Control — Regulations of the Exporting Country
a. Field Inspection of Field Crop

The seed crop is inspected regularly for diseases. Infected plants are rogued. The crop should meet requirements of the importing country.

b. Inspection of Seed Lot

The seed lot is thoroughly examined for the presence or absence of the microorganisms before export. The sample should meet the standards of the importing country.

c. Seed Treatment

The seed lot should be treated as per the requirements of the exporting country. However, treatment should conform to regulations of the importing country.

d. Phytosanitary Certificate

Phytosanitary certificates are issued by the exporting country along with the seeds as per the International Plant Protection Convention of 1951 (Rome Certificate or Phytosanitary Certificate). The validity of the certificate depends upon the test and testing methods. It has not been found as a safeguard as viewed by Neergaard, who stated:[328]

Most seed importing countries have some quarantine provisions on quarantine for seed. Many of these countries require a general plant health certificate for all or nearly all kinds of seed but do not specify any pathogen at all. Consequently, it is left entirely to the discretion of the agency of the seed exporting country to decide which seedborne pathogens should be considered and which inspection procedures should be used. As a consequence, the seed importing country has no guarantee that a seed lot, accompanied by a formally duly isssued phytosanitary certificate, has been correctly inspected for the presence of dangerous parasitic fungi, bacteria, nematodes and viruses; indeed most often no microbiological test at all has been carried out, before the certificate is signed and the seed may have been checked by visual inspection only, if at all. Needless to say that a certificate issued under such conditions is worse than useless, it is positively misleading.

3. Intermediate Quarantine

This is an international cooperative effort to lower the risk of introducing a pathogen to one country with the germplasm from another by passing this germplasm through isolation or quarantine in a third country. The pathogen in question should not pose a threat to the third country because either the crop is not grown there or the pathogen, even if it escapes, will not become established because of the environment.[326] Third-country quarantine locations are Plant Quarantine Facility, Glenn Dale, Md.; the U.S. Subtropical Horticulture Research Unit, Miami; Kew Botanical Gardens, U.K.; Royal Imperial Institute, Wageningen, The Netherlands; and IRAT at Nogent-sur Marne, France. The U.S. serves as a third country for the international exchange of coffee, tea, rubber, and cacao.[326]

F. Guidelines for Import of Germplasm[348,349]

1. Import from a country where the pathogen(s) is absent.
2. Import from a country with an efficient plant-quarantine service, so that inspection and treatment is done.
3. Obtain planting material from the safest known source within the selected country.

4.   Obtain untreated seeds so that detection of seedborne pathogens is facilitated.
5.   Obtain clean healthy-looking seeds free of any type of impurities.
6.   Obtain an official certificate of freedom from pests and diseases from the exporting country.
7.   Import the smallest possible amount of planting material; the smaller the amount the less the chance of its carrying infection. It will also simplify postentry inspection.
8.   Inspect material carefully on arrival and treat.
9.   If other precautions are not adequate, subject the material to intermediate or postentry quarantine.

VII. DISEASE RESISTANCE

The use of disease-resistant or tolerant cultivars is the most economical and efficient way of controlling diseases. However, resistance to a pathogen may not be available in all crops. Furthermore, some pathogens, such as *Cercospora sojina* (frogeye leafspot), *Heterodera glycines* (cyst nematode), and *Peronospora manshurica* (downy mildew) of soybean, exist in several races and others, such as SMV, have mild and severe strains which are expressed or repressed by temperature differences. If resistance is not available, cultivars that escape infection should be considered. The sources of resistance to soybean diseases have been summarized for selected fungal, bacterial, viral, and nematode diseases and this information may be used as guide for selecting adapted plant material.[352]

The cultivation of a cultivar may be replaced by another depending upon the degree of resistance in each. Two races, T and O, of *Drechslera maydis* cause southern blight of maize. Race T spread widely in the U.S. in 1970 and to a lesser extent in 1971. It produces a pathotoxin specific to cms-T cytoplasm of maize plants and infects the leaf, leaf sheath, husk, ear parts, and kernels. Race O, normally confined to the specific pathotoxin, primarily infects leaves. Because of the potential dangers of *D. maydis* race T, several countries have imposed legal restrictions on the importation or the planting of seeds having cms-T-cytoplasm (cytoplasm for male sterility, Texas or T). The obvious and most practical control of *D. maydis* race T is to produce the high-yielding hybrids without cms-T cytoplasm. The American seed industry and its counterpart in several other areas of the world have shifted to normal cytoplasm and to detasseling.[353]

In France, resistance is used to control Cercospora leafspot in sugar beet seed production, Verticillium wilt in alfalfa, and downy mildew in sunflower. In vegetables, resistance is used to control diseases in cucumber, spinach, bean, lettuce, melon, cabbage, pea, pepper, and tomato caused by bacteria, fungi, virus, or nematode.[354] Black-seeded cultivars of *Phaseolus vulgaris* are resistant and white-seeded are susceptible to *Rhizoctonia solani* seed infection. Extracts of black-seeded cultivars contain phenolic compounds that inhibit growth of *R. solani*.[355]

Tissue culture technique can be used for production of pathogen-free seeds. It is possible to culture soybean seedlings from the embryonic axis of a seed and grow them to maturity. These plants will produce disease-free seeds.[356]

REFERENCES

1. Agarwal, V. K., Quality seed production at Pantnagar, *Seed Sci. Technol.*, 11, 1071, 1983.
2. Gabrielson, R. L., Black leg disease of Crucifers caused by *Leptosphaeria maculans* (*Phoma lingam*) and its control, *Seed Sci. Technol.*, 11, 749, 1983.
3. Walker, J. C., Seed treatment and rainfall in relation to the control of cabbage black-leg, *U.S. Dep. Agric. Bull.*, 1029, 26, 1922.
4. Hewett, P. D., Regulating seed-borne disease by certification, in *Plant Health, the Scientific Basis for Administrative Control of Plant Diseases and Pests*, Ebbels, D. L. and King, J. E., Eds., Blackwell Scientific, Oxford, 1979, 163.
5. Gabrielson, R. L., Disease problems in cabbage seed crops, *Iowa Seed Sci.*, 2, 12, 1980.
6. Roncadori, R. W., Brooks, O. L., and Perry, C. E., Effect of field exposure on fungal invasion and deterioration of cotton seed, *Phytopathology*, 62, 1137, 1972.
7. Cunfer, B. M., The incidence of *Septoria nodorum* in wheat seed, *Phytopathology*, 68, 832, 1978.
8. Middleton, J. T. and Snyder, W. C., The production of *Ascochyta*-free pea seed in southern California, *Phytopathology*, 37, 363, 1974.
9. Leach, L. D. and MacDonald, J. D., Seed-borne *Phoma betae* as influenced by area of sugarbeet production, seed processing and fungicidal seed treatments, *J. Am. Soc. Sugar Beet Technol.*, 19, 4, 1976.
10. Baker, K. F. and Davis, L. H., Some diseases of ornamental plants in California caused by species of *Alternaria* and *Stemphylium*, *Plant Dis. Rep.*, 34, 403, 1950.
11. Butcher, C. L., Dean, L. L., and Laferriere, L., Control of halo blight of beans in Idaho, *Plant Dis. Rep.*, 52, 295, 1968.
12. Grogan, R. G. and Kimble, K. A., The role of seed contamination in the transmission of *Pseudomonas phaseolicola* in *Phaseolus vulgaris*, *Phytopathology*, 57, 28, 1967.
13. Stubbs, L. L. and O'Loughlin, G. T., Climatic elimination of mosaic spread in lettuce seed crops in the Swan Hill region of the Murray Valley, *Aust. J. Exp. Agric. Anim. Husb.*, 2, 16, 1962.
14. Kuhn, C. W. and Demski, J. W., The Relationship of Peanut Mottle Virus to Peanut Production, Res. Rep. 213, Department of Plant Pathology, University of Georgia, Athens, 1975.
15. McGee, D. C., Seed pathology: its place in modern seed production, *Plant Dis.*, 65, 638, 1981.
16. Rusch, R., Investigations into the overwintering of loose smut of oats (*Ustilago avenae* (Pers.) Jens.) and the smut reducing influence of low seed-bed temperature, *Angew. Bot.*, 31, 221, 1957.
17. Sinclair, J. B., *Compendium of Soybean Diseases*, 2nd ed., American Phytopathological Society, St. Paul, Minn., 1982, 104.
18. Agarwal, V. K., Singh, O. V., and Modgal, S. C., Influence of different doses of nitrogen and spacing on the seedborne infections of rice, *Indian Phytopathol.*, 28, 38, 1975.
19. Goldin, M. I. and Yurchenko, M. A., Method for the control of mosaic and streak in tomatoes, *Zashch. Rast. (Moscow)*, 6, 36, 1958.
20. Jones, G. H. and Seif-el-nasr, A. E. G., The influence of sowing depth and moisture on smut diseases, and prospects of a new method of control, *Ann. Appl. Biol.*, 27, 35, 1940.
21. Cralley, E. M., The effect of seeding methods on the severity of white tip of rice, *Phytopathology*, 47, 7, 1957.
22. Bisht, V. S., Sinclair, J. B., Hummel, J. W., and McClary, R. D., Effect of tillage systems on yield components and diseases of soybeans, *Phytopathology*, 72, 1134, 1982.
23. Clark, F. S., The development of an isolated area for the production of smut-free barley seed, *Agric. Inst. Rev. (Canada)*, 7, 37, 1952.
24. Kublan, A., Barley and wheat loose smut and its control, *Dtsch. Landwirtsch.*, 3, 353, 1952.
25. Oort, A. J. P., De verspreiding van de sporen vom trawestuifbrand (*Ustilago tritici*) door de lucht, *Tijdschr. Plantenziekten*, 46, 1, 1940.
26. Sinclair, J. B., Fungicide sprays for the control of seed-borne pathogens of rice, soybeans and wheat, *Seed Sci. Technol.*, 9, 697, 1981.
27. Prasartsee, C., Tenne, F. D., Ilyas, M. B., Ellis, M. A., and Sinclair, J. B., Reduction of internally seed-borne *Diaporthe phaseolorum* var. *sojae* by fungicide sprays, *Plant Dis. Rep.*, 59, 20, 1974.
28. Ellis, M. A., Foor, S. R., and Sinclair, J. B., Effect of benomyl sprays on internally-borne fungi, germination and emergence of delay harvested soybean seeds, *Phytopathol. Z.*, 85, 159, 1976.
29. Ellis, M. A., Ilyas, M. B., and Sinclair, J. B., Effect of three fungicides on internally seed-borne fungi and germination of soybean seeds, *Phytopathology*, 65, 553, 1975.
30. Ellis, M. A., Ilyas, M. B., Tenne, F. D., Sinclair, J. B., and Palm, H. L., Effect of foliar applications of benomyl on internally seed-borne fungi and pod and stem blight in soybean, *Plant Dis. Rep.*, 58, 760, 1974.
31. Ellis, M. A. and Sinclair, J. B., Effect of benomyl field sprays on internally-borne fungi, germination, and emergence of late harvested soybean seeds, *Phytopathology*, 66, 680, 1976.

32. Bolkan, H. A. and Cupertino, F. P., Control of seedborne *Phomopsis sojae* with foliar applications of fungicides, 1976, *Fungicide and Nematicide Tests,* 32, 121, 1977.
33. Tenne, F. D. and Sinclair, J. B., Control of internally seed-borne microorganisms of soybean with foliar fungicides in Puerto Rico, *Plant Dis. Rep.,* 62, 459, 1978.
34. Miller, W. A. and Roy, K. W., Effects of benomyl on the colonization of soybean leaves, pods and seeds by fungi, *Plant Dis.,* 66, 918, 1982.
35. Kmetz, K. T., Schmitthenner, A. F., and Ellett, C. W., Soybean seed decay: prevalence of infection and symptom expression caused by *Phomopsis* sp., *Diaporthe phaseolorum* var. *sojae* and *D. phaseolorum* var. *caulivora,* *Phytopathology,* 68, 838, 1978.
36. McGee, D. C. and Brandt, C. L., Effect of foliar application of benomyl on infection of soybean seeds by *Phomopsis* in relation to time of inoculation, *Plant Dis. Rep.,* 63, 675, 1979.
37. Foor, S. R. and Sinclair, J. B., Effects of fungicide sprays on soybean maturity, yield and seed quality, *Fungicide and Nematicide Tests,* 32, 122, 1977.
38. Agarwal, V. K., Singh, O. V., Thapliyal, P. N., and Malhotra, R. K., Control of purple stain disease of soybean, *Indian J. Mycol. Plant Pathol.,* 4, 1, 1974.
39. Crittenden, H. W. and Bloss, H. W., Control of *Cercospora kikuchii* and *Diaporthe phaseolorum* var. *sojae* on soybean seed, *Phytopathology,* 50(Abstr.), 570, 1960.
40. Kilpatrick, R. A., Fungi associated with the flowers, pods, and seeds of soybeans, *Phytopathology,* 47, 131, 1957.
41. Grahame, R. E., Personal communication, UniRoyal Chemical, Nautauck, N. J., 1981.
42. Jacobsen, B. J., Personal communication, Department of Plant Pathology, University of Illinois at Urbana-Champaign, 1984.
43. Cook, R. J., The effect of timed fungicide sprays on yields of winter wheat in relation to *Septoria* infection period, *Plant Pathol.,* 26, 30, 1977.
44. Jacobsen, B. J., Effect of fungicides on Septoria leaf and glume blotch, Fusarium scab, grain yield and test weight of winter wheat, *Phytopathology,* 67, 1412, 1977.
45. Tripathi, H. S., Sangam, Lal, and Agarwal, V. K., Influence of fungicidal sprays on per cent seed-borne incidence of *Fusarium moniliforme* and *Curvularia pallescens* in maize, *Pantnagar J. Res.,* 2, 104, 1977.
46. Singh, O. V., Agarwal, V. K., and Singh, R. A., Effect of fungicidal sprays on the quantum of seed-borne infection of rice, *Oryza,* 9, 103, 1972.
47. Ferrer, A., Peart, W., and Rivera, M., Control of pathogenic fungi transmitted by rice seed, *Cienc. Agropecuaria,* 3, 113, 1980.
48. Hepperly, P. R., Feliciano, C., and Sotomayor-Rios, A., Chemical control of seedborne fungi of sorghum and their association with seed quality and germination in Puerto Rico, *Plant Dis.,* 66, 902, 1982.
49. Anahosur, K. H., Chemical control of ergot of sorghum, *Indian Phytopathol.,* 32, 487, 1979.
50. Humpherson-Jones, F. M. and Maude, R. B., Control of dark leaf spot (*Alternaria brassicicola*) of *Brassica oleracea* seed production crops with foliar sprays of iprodione, *Ann. Appl. Biol.,* 100, 99, 1982.
51. Wimalajeewa, D. L. S. and Young, K. J., Studies on the levels of common and halo blight seed infection occurring in the field, *Aust. Plant Pathol.,* 8, 29, 1979.
52. Kharbanda, P. D. and Bernier, C. C., Effectiveness of seed and foliar application of fungicides to control Ascochyta blight of faba beans, *Can. J. Plant Sci.,* 59, 661, 1979.
53. Ellis, M. A. and Paschal, E. H., Effect of fungicide seed treatment on internally seed-borne fungi, germination and field emergence of pigeon pea (*Cajanus cajan*), *Seed Sci. Technol.,* 7, 75, 1979.
54. Vidhyasekaran, P. and Kandaswamy, T. K., Control of seed-borne pathogens in okra by preharvest sprays, *Indian Phytopathol.,* 33, 239, 1980.
55. Dhingra, O. D. and da Silva, J. F., Effect of weed control on the internally seedborne fungi in soybean seeds, *Plant Dis. Rep.,* 62, 513, 1978.
56. Hepperly, P. R., Kirkpatrick, B. L., and Sinclair, J. B., *Abutilon theophrasti*: wild host for three fungal parasites of soybean, *Phytopathology,* 70, 307, 1980.
57. Cerkauskas, R. F., Dhingra, O. D., Sinclair, J. B., and Asmus, G., *Amaranthus spinosus, Leonotis nepetaefolia,* and *Leonurus sibiricus*: new hosts of *Phomopsis* spp. in Brazil, *Plant Dis.,* 67, 821, 1983.
58. Schnathorst, W. C., Eradication of *Xanthomonas malvacearum* from California through sanitation, *Plant Dis. Rep.,* 50, 168, 1966.
59. Baker, K. F. and Cook, R. J., *Biological Control of Plant Pathogens,* W. H. Freeman, San Francisco, 1974, 433.
60. Weindling, K., Studies on a lethal principle effective in the parasitic action of *Trichoderma lignorum* on *Rhizoctonia solani* and other soil fungi, *Phytopathology,* 24, 1153, 1934.

61. Windels, C. E. and Kommendahl, T., Pea cultivar effect on seed treatment with *Penicillium oxalicum* in the field, *Phytopathology*, 72, 541, 1982.
62. Windels, C. E., Growth of *Penicillium oxalicum*, a biological seed treatment on pea seeds and roots in soil, *Phytopathology*, 71, 265, 1981.
63. Harman, G. E., Chet, I., and Baker, R., Factors affecting *Trichoderma hamatum* applied to seed as biological control, *Phytopathology*, 71, 569, 1981.
64. Mew, I. C. and Kommedahl, T., Biological control of seedling blight of corn by coating kernels with antagonistic microorganisms, *Phytopathology*, 58, 1395, 1968.
65. Mew, I. C. and Kommedahl, T., Interaction among microorganisms occurring naturally and applied to pericarps of corn kernels, *Plant Dis. Rep.*, 56, 861, 1972.
66. Tveit, M. and Moore, M. B., Isolates of *Chaetomium* that protect oats from *Helminthosporium victoriae*, *Phytopathology*, 44, 686, 1954.
67. Tveit, M. and Wood, R. K. S., The control of Fusarium blight in oat seedlings with antagonistic species of *Chaetomium*, *Ann. Appl. Biol.*, 43, 538, 1955.
68. Wiley, H. B. and Kommedahl, T., Biological seed treatment in sweet corn and wheat as a component of crop management, *Phytopathology*, 71, 265, 1981.
69. Marshall, D. S., Effect of *Trichoderma harzianum* seed treatment and *Rhizoctonia solani* inoculum concentration on damping-off in snapbean in acidic soils, *Plant Dis.*, 66, 788, 1982.
70. Novogrudskii, D., Beresova, E., Nachimovskaya, M., and Perviakova, M., The influence of bacterization of flax seed on the susceptibility of seedlings to infection with parasitic fungi, *C.R. Acad. Sci. U.S.S.R., N.S.*, 14, 385, 1937.
71. Price, R. D., Merriman, P. R., and Kollmorgan, J. F., The effect of seed applications of selected soil organisms on the growth and yield of cereals and carrots, *2nd Int. Congr. Plant Pathology*, St. Paul, Minn., 1973, 666.
72. Kommedahl, T. and Mew, I. C., Biocontrol of corn root infection in the field by seed treatment with antagonists, *Phytopathology*, 65, 296, 1975.
73. Henry, A. W. and Campbell, J. A., Inactivation of seed-borne plant pathogens in the soil, *Can. J. Res. Sect. C.*, 16, 331, 1938.
74. Thomas, R. C., A bacteriophage in relation to Stewart's disease of corn, *Phytopathology*, 25, 371, 1935.
75. Thomas, R. C., Additional facts regarding bacteriophage lytic to *Aplanobacter stewarti*, *Phytopathology*, 30, 602, 1940.
76. Van Winckel, A., Epidemiology of tomato mosaic, tobacco mosaic virus in tomato seed, *Agricultura (Louvain)*, 13, 721, 1965.
77. Kommedahl, T. and Windels, C. E., Evaluation of biological seed treatment for controlling rot diseases of pea, *Phytopathology*, 68, 1087, 1978.
78. Leben, C., Bacterial blight of soybean: seedling disease control, *Phytopathology*, 65, 844, 1975.
79. Merriman, P. R., Price, R. D., and Baker, K. F., The effect of inoculation of seed with antagonists of *Rhizoctonia solani* in the growth of wheat, *Aust. J. Agric. Res.*, 25, 213, 1974.
80. Merriman, P. R., Rice, R. D., Kollmorgen, J. F., Piggott, T., and Ridge, E. H., Effect of seed inoculation with *Bacillus subtilis* and *Streptomyces griseus* on the growth of cereals and carrots, *Aust. J. Agric. Res.*, 25, 219, 1974.
81. Windels, C. E. and Kommedahl, T., Factors affecting *Penicillium oxalicum* as a seed protectant against seeding blight of pea, *Phytopathology*, 68, 1656, 1978.
82. Nene, Y. L. and Thapliyal, P. N., *Fungicides in Plant Disease Control*, Oxford and IBH, New Delhi, 1979, 507.
83. Sharvelle, E. G., *Plant Disease Control*, AVI Publishing, Westport, Conn., 1979, 331.
84. Walker, J. C., *Plant Pathology*, McGraw-Hill, New York, 1969, 819.
85. Nene, Y. L. and Agarwal, V. K., *Some Important Seed-borne Diseases and their Control*, Indian Council of Agricultural Research, New Delhi, 1978, 44.
86. Keyworth, W. G. and Howell, J. S., Studies on silvering disease of redbeet, *Ann. Appl. Biol.*, 49, 173, 1961.
87. Taylor, J. D., Streptomycin seed treatment for peas and beans, *Rep. Natl. Veg. Res. Stn.*, Warwick, New Zealand, 1972.
88. Klisiewicz, J. M. and Pound, G. S., Studies on control of black rot of crucifers with antibiotics, *Phytopathology*, 50, 642, 1960.
89. Hocking, D. and Jaffar, A. A., Damping-off in pine nurseries: fungicidal control by seed pelleting, *Emp. For. Rev.*, 48, 355, 1969.
90. Walker, J. C., Onion diseases and their control, *U.S. Dep. Agric. Farmers Bull.*, 1060, 1947.
91. Byford, W. J., The incidence of sugarbeet seedling diseases and effects of seed treatment in England, *Plant Pathol.*, 21, 16, 1972.
92. Schlub, R. L. and Schmitthenner, A. F., Disinfecting soybean seeds by fumigation, *Plant Dis. Rep.*, 61, 470, 1977.

93. Ralph, W., The potential of ethylene oxide in the production of pathogen-free seed, *Seed Sci. Technol.*, 5, 567, 1977.

94. Goodey, T., *Anguillulina dipsaci* on onion seed and its control by fumigation with methyl bromide, *J. Helminthol.*, 21, 45, 1945.

95. Hague, N. G. M., Fumigation of agricultural products. XVIII. Effect of methyl bromide on the bentgrass nematode *Anguina agrostis* (Steinbuch, 1799) Filipjev 1936, and on the germination of bent grass *Agrostis tenuis*, *J. Sci. Food Agric.*, 14, 577, 1963.

96. Maude, R. B. and Presly, A. H., Neck rot (*Botrytis allii*) of bulb onions. II. Seed-borne infection in relationship to the disease in store and the effect of seed treatment, *Ann. Appl. Biol.*, 86, 181, 1977.

97. Croxall, H. E. and Hickman, C. J., The control of onion smut, *Ann. Appl. Biol.*, 40, 176, 1954.

98. Frank, Z. R., Localisation of seed-borne inocula and combined control of Aspergillus and Rhizopus rot of groundnut seedlings by seed treatment, *Is. J. Agric. Res.*, 19, 109, 1969.

99. Pommer, E. H., The systemic activity of a new fungicide of the furan carbonic acid anilide group (BAS 3191 F), *Proc. 2nd Int. Congr. Pest. Chem.*, 5, 397, 1971.

100. Leukel, R. W., Control of loose smut in oats and bunt in wheat with commercial fungicides, 1954—55, *Plant Dis. Rep.*, 39, 647, 1955.

101. Pathak, K. D., Joshi, L. M., and Renfro, B. L., Control of covered smut of oats by systemic fungicides, *Indian Phytopathol.*, 23, 693, 1970.

102. Jank, B. and Grossman, F., 2-Methyl-5-6 dihydro-4-H pyran-3-carboxylic acid, anilide: a new systemic fungicide against smut diseases, *Pest. Sci.*, 2, 43, 1971.

103. Hansing, E. D., Seed treatment with new compared with older fungicides for control of wheat, oat and sorghum smut in Kansas, 1953, *Plant Dis. Rep.*, 38, 389, 1954.

104. Gates, L. F. and Hull, R., Experiments on blackleg disease of sugarbeet seedlings, *Ann. Appl. Biol.*, 41, 541, 1954.

105. Upadhyay, J. P., Agarwal, V. K., and Mukhopadhyay, A. N., Relative efficacy of fungicidal seed treatment on emergence and seedling blight of sugarbeet, *Seed Res.*, 4, 179, 1976.

106. Sen, C., Srivastava, S. N., and Agnihotri, V. P., Seedling diseases of sugarbeet and their control, *Indian Phytopathol.*, 27, 596, 1974.

107. Maude, R. B. and Humpherson-Jones, F. M., Studies on the seedborne phases of dark leaf spot (*Alternaria brassicicola*) and grey leaf spot (*Alternaria brassicae*) of brassicas, *Ann. Appl. Biol.*, 95, 311, 1980.

108. Huber, G. A. and Gould, C. J., Cabbage seed treatment, *Phytopathology*, 39, 869, 1949.

109. Shekhawat, P. S., Jain, M. L., and Chakravarti, B. P., Detection and seed transmission of *Xanthomonas campestris* pv. *campestris* causing black rot of cabbage and cauliflower and its control by seed treatment, *Indian Phytopathol.*, 35, 442, 1982.

110. Grover, R. K. and Bansal, R. D., Seed-borne nature of *Colletotrichum capsici* in chilli seeds and its control by seed dressing fungicides, *Indian Phytopathol.*, 23, 664, 1970.

111. Vidhyasekaran, P. and Thiagarajan, C. P., Seed-borne transmission of *Fusarium oxysporum* in chilli, *Indian Phytopathol.*, 34, 211, 1981.

112. Dharam, V. and Grewal, J. S., Efficacy of different fungicides. III. Seed disinfection in relation to damping-off of chillies (*Capsicum annuum* L.), *Indian Phytopathol.*, 14, 10, 1961.

113. Reddy, M. V., Singh, K. B., and Nene, Y. L., Further studies on Calixin M in the control of seed-borne infection of Ascochyta blight in chickpea, *Int. Chickpea Newsl.*, 6, 18, 1982.

114. Zachos, D. G., Panagopulos, C. G., and MaKris, S. A., Researches on the biology, epidemiology and the control of anthracnose of chickpea, *Ann. Inst. Phytopathol. Benaki*, 5, 167, 1963.

115. Lukashevich, A. I., Control measures against ascochytosis of chickpea, *J. Agric. Sci. (Moscow)*, 5, 131, 1958.

116. Khachatryan, M. S., Seed transmission of ascochytosis infection in chickpea and the effectiveness of treatment, *Sb. Nauchn. Tr. Nauchno Issled. Zemledel. Armyarskoi*, 2, 147, 1961.

117. Kaiser, W. J., Okhovat, M., and Mossahebi, G. H., Effect of seed treatment fungicides on control of *Ascochyta rabiei* in chickpea seed infected with the pathogen, *Plant Dis. Rep.*, 57, 742, 1973.

118. Nene, Y. L., Siddiqui, I. A., and Kharbanda, P. D., Control of stemgall of coriander by fungicides, *Mycopathol. Mycol. Appl.*, 29, 142, 1966.

119. Grewal, J. S. and Dharam, V., Efficacy of different fungicides. VI. Field trials for the control of stem rot of jute, *Indian Phytopathol.*, 16, 99, 1963.

120. Agarwal, V. K. and Singh, O. V., Seed-borne fungi of jute and their control, *Indian Phytopathol.*, 27, 651, 1974.

121. Hildebrand, A. A., Seedborne diseases of soybean and their control, *Proc. Can. Phytopathol. Soc.*, 12, 18, 1944.

122. Nene, Y. L., Agarwal, V. K., and Srivastava, S. S. L., Influence of fungicidal seed treatment on the emergence and nodulation of soybean, *Pesticides*, 3, 26, 1969.

123. Singh, O. V., Agarwal, V. K., and Nene, Y. L., Influence of fungicidal seed treatment on the myco-flora of stored soybean seed and seedling emergence, *Indian J. Agric. Sci.*, 43, 820, 1973.
124. Al-Beldawi, A. S. and Welleed, B. C., Chemical control of *Rhizoctonia solani* Kühn on cotton seed-lings, *Phytopathol. Mediterr.*, 12, 87, 1973.
125. Kotasthane, S. R. and Agarwal, S. C., Efficacy of four seed dressing fungicides in controlling black arm of cotton, *PANS*, 16, 334, 1970.
126. Paulus, A. O., Nelson, J., Dewolfe, T., House, J., and Shibuya, F., Non-mercury fungicides for control of seedling disease of cotton, *Calif. Agric.*, 27, 9, 1973.
127. Nikolov, G., Apron 35SD an effective preparation in the control of downy mildew of sunflower, *Rastit. Zasht.*, 29, 40, 1981.
128. Tollenaar, H. and Bleiholder, H., Distribution of the mycelium of *Sclerotinia sclerotiorum* in sun-flower seed, *Agric. Tec. Mex.*, 31, 44, 1971.
129. Sackston, W. E., *Botrytis cinerea* and *Sclerotinia sclerotiorum* in seed of safflower and sunflower, *Plant Dis. Rep.*, 44, 664, 1960.
130. Singh, O. V. and Agarwal, V. K., Influence of fungicidal seed treatment on emergence and seedborne mycoflora of sunflower, *Labdev, Part B*, 11, 56, 1973.
131. Kingsland, G. C., Barley leaf stripe control by Vitavax, *Phytopathology*, 60, 584, 1970.
132. Grewal, J. S. and Dharam, V., Efficacy of different fungicides. I. Seed disinfection in relation to stem rot of jute and stripe disease of barley, *Indian Phytopathol.*, 11, 175, 1958.
133. Moseman, J. G., Fungicidal Control of Smut Diseases of Cereals, Circ. 42, U.S. Department of Agriculture, Washington, D. C., 42, 1968.
134. Grewal, J. S. and Dharam, V., Efficacy of different fungicides. VIII. Field trials for the control of covered smut of barley (*Ustilago hordei* (Pers.) Lager.), *Indian Phytopathol.*, 17, 162, 1964.
135. von Schmeling, B. and Kulka, M., Systemic fungicidal activity of 1,4-oxathiin derivatives, *Science*, 152, 659, 1966.
136. Pommer, E. H. and Kradel, J., 2,5-dimethyl-Furane-3-carboxylic acid anilide (BAS 3191 F) a new active ingredient for the control of seed-borne fungus disease in cereal, in 7th Int. Congr. Plant Protection, Paris, 1970, 409.
137. Kovacikova, E., Seed treatment of lentil and pea against some fungal diseases, *Ochr. Rost.*, 6, 117, 1970.
138. Agarwal, S. C. and Kotasthane, S. R., Linseed rust and its control, *Telhan Patrika*, 2, 18, 1970.
139. Agarwal, V. K. and Singh, O. V., Seed-borne fungi of linseed and their response to seed treatment, *Seed Res.*, 3, 26, 1975.
140. Devash, Y., Okon, Y., and Henis, Y., Survival of *Pseudomonas tomato* in soil and seeds, *Phytopathol. Z.*, 99, 175, 1980.
141. Cole, J. S., Some control measures for anthracnose disease of tobacco, *Ann. Appl. Biol.*, 45, 542, 1957.
142. Dharam, V., Mathur, S. B., and Neergaard, P., Efficacy of certain fungicides against seed-borne infection of stackburn disease of rice caused by *Trichoconis padwickii*, *Indian Phytopathol.*, 24, 343, 1971.
143. Agarwal, V. K. and Singh, O. V., Studies on the detection of seed-borne fungi of rice and their control, *Bull. Grain Technol.*, 11, 189, 1973.
144. Misra, A. P. and Singh, T. B., Effect of some copper and organic fungicide on the viability of paddy seeds, *Indian Phytopathol.*, 22, 264, 1969.
145. Dharam, V., Mathur, S. B., and Neergaard, P., Control of seed-borne infection of *Drechslera* spp. on barley, rice and oats with Dithane M-45, *Indian Phytopathol.*, 23, 570, 1970.
146. Bedi, K. S., Paracer, C. S., and Chohan, J. S., Tackle foot rot to save the rice crop, *Indian Farm.*, 8, 17, 1958.
147. Ribeiro, A. S., Seed treatment of irrigated rice for control of the spread of *Aphelenchoides besseyi*, Rio Grande do sul, *Empressa Bras. Pesqui. Agropecu.*, (Pelotas, Brazil), 138, 1977.
148. Muthusamy, M. and Narayanasamy, P., Seed transmission of pearlmillet downy mildew and its con-trol, *Indian Phytopathol.*, 34, 418, 1981.
149. Cox, R. S., Stem anthracnose of lima beans and its control, *Phytopathology*, 38, 7, 1948.
150. Taylor, J. D. and Dudley, C. L., Seed treatment for the control of halo-blight of beans (*Pseudomonas phaseolicola*), *Ann. Appl. Biol.*, 85, 223, 1977.
151. Vlakhov, S., Kutova, I., and Koleva, P., Action of antibiotics against some bacterioses, *Rostenievdni Nauki*, 11, 123, 1974.
152. Anderson, A. L. and Dezcevew, D. J., Seed treatment studies for damping-off control in garden and canning beans, *Rep. Prog. Q. Bull. Mich. Agric. Exp. Stn.*, 34, 357, 1952.
153. Kirik, N. N., Influence of the depth of mycelial penetration to the causal agent of ascochytosis into pea seeds on the effectiveness of treatment, *Mikol. Fitopatol.*, 4, 419, 1970.
154. Yoshii, K., Seed treatment of pea with benomyl to control *Mycosphaerella pinodes*, *Fitopatologia*, 10, 41, 1975.

155. Miller, M. W. and de Whalley, C. V., The use of metalaxayl seed treatments to control pea downy mildew, *Proc. Br. Crop Protection Conf.* Vol. 1, Brighton, England, 341, 1981.

156. Taylor, J. D. and Dye, D. W., Evaluation of streptomycin seed treatments for the control of bacterial blight of peas (*Pseudomonas pisi* Sackett 1916), *N. Z. J. Agric. Res.*, 19, 91, 1976.

157. Crosier, W., Chemical control of seed-borne fungi during germination on testing of peas and sweet-corn, *Phytopathology*, 36, 92, 1946.

158. Nene, Y. L. and Agarwal, V. K., Influence of fungicidal seed treatment on emergence and yield of pea var. Bridger, *Pesticides*, 3, 15, 1969.

159. Tisdale, W. B., Brooks, A. N., and Townsend, G. R., Dust treatments for vegetable seed, *Bull. Fla. Agric. Exp. Stn.*, 413, 32, 1945.

160. Shukla, B. N. and Singh, B. P., Effect of fungicidal seed treatment on Macrophomina root rot of sesame (*Sesamum indicum*), *Indian J. Mycol. Plant Pathol.*, 2, 208, 1973.

161. Venugopal, M. N. and Safeeulla, K. M., Chemical control of the downy mildew of pearlmillet, sorghum and maize, *Indian J. Agric. Sci.*, 48, 537, 1978.

162. Grewal, J. S. and Dharam, V., Efficacy of different fungicides. IV. Field trials for the control of grain smut of jowar *Sphacelotheca sorghi* (Link) Clinton, *Indian Phytopathol.*, 14, 213, 1961.

163. Webster, O. J. and Leukel, R. W., Sorghum seed treatment tests in 1958, *Plant Dis. Rep.*, 43, 348, 1959.

164. Leukel, R. W., Spergon as a seed disinfectant, *Plant Dis. Rep.*, 26, 93, 1942.

165. Hansing, E. D., Seed treatment with new compared with older fungicides for control of wheat, oat and sorghum smut in Kansas, 1953, *Plant Dis. Rep.*, 38, 389, 1954.

166. Hansing, E. D. and Melchers, L. E., Standard and new fungicides for the control of covered smut of sorghum and their effect on stand, *Phytopathology*, 34, 1034, 1944.

167. Agarwal, V. K., Verma, H. S., and Singh, O. V., Treatment of sorghum seeds to control seed-borne fungi and improve emergence. *Bull. Grain Technol.*, 15, 118, 1977.

168. Lee, D. H., Control of seed-borne infection of *Ustilago nuda* and *Pyrenophora graminea* on barley, *Korean J. Mycol.*, 8, 89, 1980.

169. Machacek, J. F., Cooperative seed treatment trials, 1953, *Plant Dis. Rep.*, 38, 169, 1954.

170. Siljes, I. and Halbauer, V., Possibilities of treating wheat seed with systemic fungicide, *Agron. Glas.*, 33, 447, 1974.

171. Bateman, G. L., The efficacy of two organomercury compounds in controlling seedborne *Septoria nodorum* on winter wheat, *Ann. Appl. Biol.*, 79, 307, 1975.

172. Sharma, R. C., Joshi, L. M., and Pathak, K. D., Systemic fungicides for the control of hill bunt of wheat, *Indian Phytopathol.*, 24, 604, 1971.

173. Leukel, R. W., Cooperative seed treatment on small grains in 1951, *Plant Dis. Rep.*, 35, 445, 1951.

174. Grewal, J. S., Joshi, P. C., Pathak, K. D., and Mathur, S. B., Relative efficacy of some new fungicides for the control of hill bunt of wheat, *Indian Phytopathol.*, 18, 94, 1965.

175. More, K. J. and Kuiper, J., New treatments for bunt of wheat, *Agric. Gaz. N. S. W.*, 85, 16, 1974.

176. Line, R. F., Chemical control of flag smut of wheat, *Plant Dis. Rep.*, 56, 636, 1972.

177. Chatrath, M. S., Renfro, B. L., Nene, Y. L., Grover, R. K., Roy, M. R., Singh, D. V., and Gandhi, S. M., Control of loose smut of wheat with systemic fungicides, *Indian Phytopathol.*, 22, 184, 1969.

178. Agarwal, V. K., Agarwal, M., Verma, H. S., and Gupta, R. K., Studies on loose smut of wheat. III. Influence of infection on plant morphology and control through seed treatment with carboxin, *Seed Res.*, 10, 79, 1982.

179. Tyagi, P. D., Singh, M., and Chauhan, M. S., Comparative efficacy of some systemic fungicides for controlling loose smut of wheat, *Pesticides*, 10, 26, 1976.

180. Sharma, J. K., Aujla, S. S., Sharma, Y. R., and Chauhan, J. S., Systemic fungicides for the control of loose smut of wheat, *Pesticides*, 12, 30, 1978.

181. Agarwal, V. K., Singh, A., and Verma, H. S., A note on emergence of wheat seed treated with different seed dressing fungicides prior to storage, *Seed Res.*, 4, 194, 1976.

182. Warren, H. L. and Nicholson, R. L., Kernel infection, seedling blight and wilt of maize caused by *Colletotrichum graminicola*, *Phytopathology*, 65, 620, 1975.

183. Lim, S. M. and Kinsey, J. G., Seed treatment of corn infected with *Helminthosporium maydis* race T., *Plant Dis. Rep.*, 57, 344, 1973.

184. Simpson, W. R. and Fenwick, H. S., Suppression of corn head smut with carboxin seed treatments, *Plant Dis. Rep.*, 55, 501, 1971.

185. Berger, R. D. and Wolf, E. A., Control of seed-borne and soil-borne mycoses of "Florida sweet" corn by seed treatment, *Plant Dis. Rep.*, 58, 922, 1974.

186. Hoppe, P. E., Comparison of certain mercury and non-metallic dusts for corn seed treatment, *Phytopathology*, 33, 602, 1943.

187. Verma, H. S. and Agarwal, V. K., A note on the influence of prestorage fungicidal seed treatment on emergence of maize seeds, *Bull. Grain. Technol.*, 19, 57, 1982.

188. Beaumont, A., Cleary, J. P., and Bant, J. H., Control of damping-off of zinnias caused by *Alternaria zinniae, Plant Pathol.*, 7, 52, 1958.
189. Strider, D. L., Eradication of *Xanthomonas nigromaculans* f. sp. *zinniae* in zinnia seed with sodium hypochlorite, *Plant Dis. Rep.*, 63, 873, 1979.
190. Demski, J. W., Tobacco mosaic virus is seed-borne in pimiento peppers, *Plant Dis.*, 65, 723, 1981.
191. Alexander, L. J., Inactivation of tobacco mosaic virus from tomato seed, *Phytopathology*, 50, 627, 1960.
192. Taylor, R. H., Grogan, R. G., and Kimble, K. A., Transmission of tobacco mosaic virus in tomato seed, *Phytopathology*, 51, 837, 1961.
193. Gooding, G. V., Jr., Inactivation of tobacco mosaic virus on tomato seed with trisodium orthophosphate and sodium hypochlorite, *Plant Dis. Rep.*, 59, 770, 1975.
194. Sharma, S. R. and Varma, A., Cure of seed transmitted cowpea banding mosaic disease, *Phytopathol. Z.*, 83, 144, 1975.
195. Walkey, D. G. A. and Dance, M. C., High temperature inactivation of seed borne lettuce mosaic virus, *Plant Dis. Rep.*, 63, 125, 1979.
196. Kadian, O. P., Effects of some chemicals and heat on seed transmission of urdbean leaf crinkle virus, in 3rd Int. Symp. Plant Pathology, New Delhi, 1981, 166.
197. Todd, E. H. and Atkins, J. C., White tip disease of rice. II. Seed treatment studies, *Phytopathology*, 49, 184, 1959.
198. Fukano, H., Ecological studies on white tip disease of rice plant caused by *Aphelenchoides besseyi* Christie and its control, *Fukuoka Agric. Exp. Stn. Bull.*, 18, 108, 1962.
199. Reddy, M. V., Calixin M — an effective fungicide for eradication of *Ascochyta rabiei* in chickpea seed, *Int. Chickpea News Lett.*, 3, 12, 1980.
200. Haware, M. P., Nene, Y. L., and Rajeshwari, R., Eradication of *Fusarium oxysporum* f. sp. *ciceri* transmitted in chickpea seed, *Phytopathology*, 68, 1364, 1978.
201. Agarwal, V. K., Assessment of seed-borne infection and treatment of wheat seeds for the control of loose smut, *Seed Sci. Technol.*, 9, 725, 1981.
202. Sanderson, F. R. and Hampton, J. G., Role of perfect states in the epidemiology of the common Septoria diseases of wheat, *N.Z. J. Agric. Res.*, 21, 277, 1978.
203. Klitgard, K. and Jørgensen, J., The correlation between the germination percentage determined in the laboratory and the field with samples of seeds of winter wheat infected with *Septoria nodorum, Statsfrøkont. Beret.*, 102, 85, 1973.
204. Hermansen, J. E. and Jørgensen, J., Historical aspects of the control of seed-borne cereal diseases in Denmark, *Seed Sci. Technol.*, 11, 1005, 1983.
205. Jørgensen, J., Disease testing of barley seed and application of test results in Denmark, *Seed Sci. Technol.*, 11, 615, 1983.
206. Anon., *Annual Report of the Swedish Seed Testing and Certification Institute 1977/79.* Maddelande fran statens centrala Frokontrollanstalt, Stockholm, 54, 85, 1979.
207. Ralph, W., Pelleting seed with bactericides — the effect of streptomycin on seed-borne halo blight of French-bean, *Seed Sci. Technol.*, 4, 325, 1976.
208. Ellis, M. A. and Paschal, E. H., Transfer of technology in seed pathology of tropical legumes, in *Seed Pathology — Problems and Progress,* Yorinori, J. T., Sinclair, J. B., Mehta, Y. R., and Mohan, S. K., Eds., Fundação Instituto Agronômico do Paraná, IAPAR, Londrina, Brazil, 1979, 190.
209. Cook, A. A., Larson, R. H., and Walker, J. C., Relation of the black rot pathogen to cabbage seed, *Phytopathology*, 42, 316, 1952.
210. Patel, P. N., Trivedi, B. M., Rekhi, S. S., Town, P. A., and Rao, Y. P., Black rot and stump rot in cauliflower seed crops in India, *FAO Plant Prot. Bull.*, 18, 136, 1970.
211. Harrower, K. M., Tolerance of *Leptosphaeria nodorum* to an organomercurial compound, *Trans. Br. Mycol. Soc.*, 66, 523, 1976.
212. Noble, M. and Macgarvie, Q. D., Resistance to mercury of *Pyrenophora avenae* in Scottish seed oats, *Plant Pathol.*, 15, 23, 1966.
213. Kuiper, J., Failure of hexachlorobenzene to control common bunt of wheat, *Nature (London)*, 206, 1219, 1965.
214. Old, K. M., Mercury tolerant *Pyrenophora avenae* in seed oats, *Trans. Br. Mycol. Soc.*, 51, 525, 1968.
215. Ellis, M. A. and Sinclair, J. B., Uptake and translocation of streptomycin by seedlings, *Plant Dis. Rep.*, 58, 534, 1974.
216. Humaydan, H. S., Harman, G. E., Nedrow, B. L., and DiNitto, L. V., Eradication of *Xanthomonas campestris*, the causal agent of blackrot from Brassica seeds with antibiotics and sodium hypochlorite, *Phytopathology*, 70, 127, 1980.
217. Anahosur, K. H. and Patil, S. H., Chemical control of sorghum downy mildew in India, *Plant Dis.*, 64, 1004, 1980.

218. Ryker, T. C., Seed coloration, in *Proc. 1959 Short Course for Seedsmen,* University of Florida, 1959, 123.

219. Meyer, H. and Mayer, A. M., Permeation of dry seeds with chemicals: use of dichloromethane, *Science,* 171, 683, 1971.

220. Elden, M., Mayer, A. M., and Poljakoff-Mayer, A., Permeation of dry lettuce seeds with acetic anhydride and with amino acids, using dichloromethane, *Seed Sci. Technol.,* 2, 317, 1974.

221. Royce, D. J., Ellis, M. A., and Sinclair, J. B., Movement of penicillin into soybean seeds using dichloromethane, *Phytopathology,* 65, 1319, 1975.

222. Ralph, W., A note on antibiotic permeation of seed with dichloromethane, *Seed Sci. Technol.,* 5, 575, 1977.

223. Ellis, M. A., Foor, S. R., and Sinclair, J. B., Dichloromethane: nonaqueous vehicle for systemic fungicides in soybean seeds, *Phytopathology,* 66, 1249, 1976.

224. Papavizas, G. C. and Lewis, J. A., Acetone infusion of pyroxychlor into soybean seed for the control of *Phytophthora megasperma* var. *sojae, Plant Dis. Rep.,* 60, 484, 1976.

225. Hepperly, P. R. and Sinclair, J. B., Aqueous polyethylene glycol solutions for treating soybean seeds with antibiotics, *Seed Sci. Technol.,* 5, 727, 1977.

226. Tao, K. L., Khan, A. A., Harman, G. E., and Eckenrode, C. J., Practical significance of the application of chemical in organic solvents to dry seeds, *J. Am. Soc. Hortic. Sci.,* 99, 217, 1974.

227. Heydecker, W., Higgins, J., and Turner, Y. J., Invirogration of seeds, *Seed Sci. Technol.,* 3, 881, 1975.

228. Browing, E., *Toxicology and Metabolism of Industrial Solvents,* Elsevier, Amsterdam, 1965.

229. Sinclair, J. B., Soybean seed pathology, in *Seed Pathology — Problems and Progress,* Yorinori, J. T., Sinclair, J. B., Mehta, Y. R., and Mohan, S. K., Eds., Fundação Instituto Agronômico do Paraná, IAPAR, Londrina, Brazil, 1979, 161.

230. Shortt, B. J. and Sinclair, J. B., Efficacy of polyethylene glycol and organic solvents for infusing fungicides into soybean seeds, *Phytopathology,* 70, 971, 1980.

231. Muchovej, J. J. and Dhingra, O. D., Benzene and ethanol for treatment of soybean seeds with systemic fungicides, *Seed Sci. Technol.,* 7, 449, 1979.

232. Damicone, J. P., Cooley, D. R., and Manning, W. J., Elimination of Fusaria from asparagus seed, *Phytopathology,* 70, 461, 1980.

233. Oneill, N. R., Papavizas, G. C., and Lewis, J. A., Infusion and translocation of systemic fungicides applied to seeds in acetone, *Phytopathology,* 69, 690, 1979.

234. Kraft, J. M., Comparison of acetone infusion to slurry application of fungicides to pea seed, *Phytopathology,* 71, 232, 1981.

235. Kraft, J. M., Field and greenhouse studies on pea seed treatments, *Plant Dis.,* 66, 798, 1982.

236. Muchovej, J. J. and Dhingra, O. D., Acetone, benzene and ethanol for treating *Phaseolus* bean seeds in the dry state with systemic fungicides, *Seed Sci. Technol.,* 8, 351, 1980.

237. Agarwal, V. K. and Sinclair, J. B., Seed dressings and *Rhizobium* inoculum, in *Soybean Seed Quality and Stand Establishment* (INTSOY Ser. No. 22), Sinclair, J. B. and Jacobs, J. A., Eds., College of Agriculture, University of Illinois at Urbana-Champaign, 1981, 127.

238. Curley, R. L. and Burton, J. C., Compatibility of *Rhizobium japonicum* with chemical seed protectants, *Agron. J.,* 67, 807, 1975.

239. Furtode, A., Effect of Systemic Fungicides on Soil Microbial Population and Nodulation by *Rhizobium* spp., Thesis Abstr., College of Agriculture, Dharwar, India, 3, 286, 1977.

240. Brinkerhoff, L. A., Fink, G., Kortsen, R. A., and Swift, D., Further studies on the effect of chemical seed treatments on nodulation of legumes, *Plant Dis. Rep.,* 38, 393, 1954.

241. Kis, G., Papp, I., Bakondizamori, E., and Gartner Banfalvi, A., Study of soybean seed dressing with fungicides combined with *Rhizobium* inoculation, *Novenytermeles,* 26, 147, 1977.

242. Wankhede, V. K. and Bhide, V. P., Compatibility of different fungicides with *Rhizobium japonicum, Hind. Antibiot. Bull.,* 14, 131, 1972.

243. Tu, C. M., Effects of pesticide seed treatments on *Rhizobium japonicum* and its symbiotic relationship with soybean, *Bull. Environ. Contam. Toxicol.,* 18, 190, 1977.

244. Ganacharya, N. M., Effect of fungicidal seed treatment on emergence, nodulation and grain yield of soybean in Marathwada, *J. Maharashtra Agric. Univ.,* 4, 112, 1979.

245. Hamdi, Y. A., Moharram, A. A., and Lofti, M., Effect of certain fungicides on some Rhizobia legume symbiotic systems, *Zentralbl. Bakeriol. Parasitenkd. Infektionskr. Hyg. Zweite Naturwiss. Abt. Allg. Landwirtsch. Tech. NMkro Ea,* 3—4, 363, 1974.

246. Batalova, T. S., Zinovev, L. S., Kiselev, I. I., Kikhanina, K. A., and Masiutina, V. A., Compatibility of the bacterial fertilizer nitrogen treatment and dressing of legume seed fungicides, *Khim. Sel'sk. Khoz.,* 15, 37, 1977.

247. Nery, M. and Dobereiner, J., Effect of preemergent fungicides on nodulation and N_2 fixation in soybean, in Anais do Decimo Quinto Congr. Brasileiro de Ciencia do Solo, Sao Paolo, Brazil, 1976, 177.

248. Koehler, B., Results of uniform seed treatment tests on soybeans, *Plant Dis. Rep.*, Suppl. 145, 76, 1943.

249. Maggione, C. S. and Lamsanchez, A., Effect of seed treatment with thiabendazol, alone and in combination with captan, on germination and nodulation of soybean (*Glycine max* (L.) Merrill), *Cientifica*, 4, 107, 1976.

250. Backman, P. A., Effects of seed treatment fungicides on *Rhizobium* inoculants, *Highlights Agric. Res.*, 25, 14, 1978.

251. Burton, J. C., Problems in obtaining adequate inoculation of soybeans, in *World Soybean Research*, Hill, L. D., Ed., Interstate Printers, Danville, Ill., 1976, 170.

252. Staphorst, J. L. and Strijdom, B. W., Effects on Rhizobia of fungicides applied to legume seed, *Phytophylactica*, 8, 47, 1976.

253. Odeyemi, O., Resistance of *Rhizobium* to thiram, Spergon and Phygon, *Diss. Abstr. Int. B*, 38, 993, 1977.

254. Bewley, W. F. and Corbett, W., The control of cucumber and tomato mosaic disease in glasshouse by the use of clean seed, *Ann. Appl. Biol.*, 17, 260, 1930.

255. MacDonald, J. D. and Leach, L. D., The association of *Fusarium oxysporum* f. sp. *betae* with nonprocessed and processed sugarbeet seed, *Phytopathology*, 66, 868, 1976.

256. Tomlinson, J. A. and Faithfull, E. M., *23rd Annual Report for 1972*, National Vegetable Research Institute, Wellesbourne, England, 1973, 139.

257. Ryder, E. J. and Johnson, A. S., A method for indexing lettuce seeds for seed-borne lettuce mosaic virus by air-stream separation of light from heavy seeds, *Plant Dis. Rep.*, 58, 1037, 1974.

258. Phatak, H. C. and Summanwar, A. S., Detection of plant viruses in seeds and seed stocks, *Proc. Int. Seed. Test. Assoc.*, 32, 625, 1967.

259. Stevenson, W. R. and Hagedorn, D. J., Effect of seed size and condition on transmission of pea seed-borne mosaic virus, *Phytopathology*, 60, 1148, 1970.

260. Thyr, B. D., Webb, R. E., Jaworski, C. A., and Ratcliffe, T. J., Tomato bacterial canker: control by seed treatment, *Plant Dis. Rep.*, 57, 974, 1973.

261. Proctor, C. H. and Fry, P. R., Seed transmission of tobacco mosaic virus in tomato, *N.Z. J. Agric. Res.*, 8, 367, 1965.

262. Milenko, Y. F., Cleaning of white rot sclerotia from sunflower seeds, *Sel. Seed-Gr. (Moscow)*, 29, 73, 1964.

263. Hepperly, P. R. and Sinclair, J. B., A glycerin and polyethyleneglycol solution for separating healthy and diseased soybean seeds, *Seed Sci. Technol.*, 11, 125, 1982.

264. Brinkerhoff, L. A. and Hunter, R. E., Internally infected seed as a source of inoculum for the primary cycle of bacterial blight of cotton, *Phytopathology*, 53, 1397, 1963.

265. Baker, K. F., Thermotherapy of planting material, *Phytopathology*, 52, 1244, 1962.

266. Watson, R. D., Coltrin, L., and Robinson, R., The evaluation of materials for heat treatment of peas and beans, *Plant Dis. Rep.*, 35, 542, 1951.

267. Zinnen, T. M. and Sinclair, J. B., Thermotherapy of soybean seeds to control seedborne fungi, *Phytopathology*, 72, 831, 1982.

268. Luthra, J. C., Solar energy treatment of wheat loose smut *Ustilago tritici* (Pers.) Rostr., *Indian Phytopathol.*, 6, 49, 1953.

269. Smith, P. R., Seed transmission of *Itersonilia pastinacae* in parsnip and its elimination by a steam-air treatment, *Aust. J. Exp. Agric. Anim. Husb.*, 6, 441, 1966.

270. Krasnova, M. V., The effect of some physical factors on the causal agents of bacterioses in soybean seeds, *J. Microbiol. (Kiev)*, 25, 50, 1963.

271. Hankin, I. and Shands, D. C., Microwave treatment of tobacco seed to eliminate bacteria on the seed surface, *Phytopathology*, 67, 794, 1977.

272. Halliwell, R. S. and Langston, R., Effects of gamma irradiation on symptom expression of barley stripe mosaic virus disease and on two viruses *in vivo*, *Phytopathology*, 55, 1039, 1965.

273. Megahed, E. S. and Moore, A., Inactivation of necrotic ringspot and prune dwarf viruses in seeds of some *Prunus* spp., *Phytopathology*, 59, 1758, 1969.

274. Bridge, J., Bos, W. S., Page, L. T., and McDonald, D., The biology and possible importance of *Aphelenchoides arachidis*, a seed-borne ectoparasitic nematode of ground-nut from northern Nigeria, *Nematologica*, 23, 253, 1977.

275. Sharvelle, E. G., *Plant Disease Control*, AVI Publishing, Westport, Conn., 1979, 331.

276. Leben, C., Control of *Pseudomonas lachrymans* in cucumber seed by a temperature-relative humidity method, *Phytopathology*, 71, 235, 1981.

277. Srivastava, D. N. and Rao, Y. P., Epidemiology and control of bacterial blight of guar (*Cyamopsis tetra-gonoloba* (L.) Taub.) *Bull. Indian Phytopathol. Soc.*, 6, 1, 1970.

278. Thorne, G., *Principles of Nematology*, McGraw Hill, New York, 1961, 553.

279. Ohata, K., Serizawa, S., Azegami, K., and Shirata, A., Possibility of seed transmission of *Xantho-monas campestris* pv. *vitians*, the pathogen of bacterial spot of lettuce, *Bull. Nat. Inst. Agric. Sci.,* C, 36, 81, 1982.

280. McIntyre, J. L., Sands, D. C., and Taylor, G. S., Overwintering, seed disinfestation, and pathogen-icity studies of the tobacco hollow stalk pathogen, *Erwinia carotovora* var. *carotovora, Phytopath-ology,* 68, 435, 1978.

281. Mohanty, N. N., Control of Udbatta disease of rice, *Proc. Indian Nat. Sci. Acad.,* B37, 432, 1971.

282. Todd, E. H. and Atkins, J. G., White tip disease of rice. II. Seed treatment studies, *Phytopathology,* 49, 184, 1959.

283. Thakur, D. P. and Kanwar, Z. S., Internal seed-borne infection and heat therapy in relation to downy mildew of *Pennisetum typhoides* Stapf. and Hubb., *Sci. Cult.,* 43, 433, 1977.

284. Baker, K. F., Bacterial fasciation disease of ornamental plants in California, *Plant Dis. Rep.,* 34, 121, 1950.

285. Reddick, D. and Stewart, V. B., Transmission of the virus of bean mosaic in seed and observations on thermal death point of seed and virus, *Phytopathology,* 9, 445, 1919.

286. Owusu, G. K., Crowley, N. C., and Francki, R. I. B., Studies of the seed transmission of tobacco ringspot virus, *Ann. Appl. Biol.,* 61, 195, 1968.

287. Rohloff, I., Trials for inactivation of lettuce mosaic virus in seeds, *Gartenbauwissenschaft,* 28, 19, 1963.

288. Timian, R. G., Heat treatments fail to inactivate barley stripe mosaic virus in seed, *Plant Dis. Rep.,* 49, 696, 1965.

289. Vovk, A. M., Inactivation of tobacco mosaic virus in tomato seed at different storage times, *Tr. Inst. Genet. Akad. Nauk. U.S.S.R,* 28, 269, 1961.

290. Howles, R., Inactivation of tomato mosaic virus in tomato seeds, *Plant Pathol.,* 10, 160, 1961.

291. Broadbent, L. H., The epidemiology of tomato mosaic. XI. Seed transmission of TMV, *Ann. Appl. Biol.,* 56, 177, 1965.

292. Laterrot, H. and Pecaut, P., Tomato seed production. I. Rapid cleansing using pectolytic enzymes. II. Decreasing the content of tobacco mosaic virus by dry heat treatment, *Ann. Epiphyt.,* 16, 163, 1965.

293. Laterrot, H. and Pecaut, P., Incidence du traitement thermique des semences de tomate sur la trans-mission du virus de la mosaique du tabac, *Ann. Epiphyt.,* 19, 159, 1968.

294. Van Dorst, H. J. M., Investigation into cucumber virus 2, *Groenten Fruit,* 22, 1519, 1967.

295. Fletcher, J. T., George, A. J., and Green, D. E., Cucumber green mottle virus, its effect on yield and its control in the Lea Valley, England, *Plant Pathol.,* 18, 16, 1969.

296. Verma, V. S., Effect of heat on seed transmission of mosaic disease of cowpea (*Vigna sinensis* Savi), *Acta Microbiol. Pol. Ser. B.,* 3, 163, 1971.

297. Sharma, Y. R. and Chohan, J. S., Control by thermotherapy of seed-borne vegetable marrow mosaic virus, *FAO Plant Prot. Bull.,* 19, 86, 1971.

298. Bloom, J. R., Lethal effect of temperature extremes on *Anguina tritici, Phytopathology,* 53, 347, 1963.

299. Srinivasan, M. C., Neergaard, P., and Mathur, S. B., A technique for detection of *Xanthomonas campestris* in routine seed health testing of crucifers, *Seed Sci. Technol.,* 1, 853, 1973.

300. Ralph, W., Problems in testing and control of seed-borne bacterial pathogens: a critical review, *Seed Sci. Technol.,* 5, 735, 1977.

301. Cafati, C. R. and Saettler, A. W., Transmission of *Xanthomonas phaseoli* in seed of resistant and susceptible *Phaseolus* genotypes, *Phytopathology,* 70, 638, 1980.

302. Rennie, W. J. and Seaton, R. D., Loose smut of barley. The embryo test as a means of assessing loose smut infection in seed stocks, *Seed Sci. Technol.,* 3, 697, 1975.

303. Doling, D. A., Loose smut in wheat and barley, *Agriculture (London),* 73, 523, 1966.

304. Neergaard, P., *Seed Pathology,* Vol. 1 and 2, Macmillan, London, 1977, 1187.

305. Hewett, P. D., The field behaviour of seed-borne *Ascochyta fabae* and disease control in field beans, *Ann. Appl. Biol.,* 74, 287, 1973.

306. Butcher, C. L., Dean, L. L., and Guthrie, J. W., Effectiveness of halo blight control in Idaho bean seed crops, *Plant Dis. Rep.,* 53, 894, 1969.

307. Copeland, L. O. and Adams, M. W., An improved seed program for maintaining disease-free seed of field beans (*Phaseolus vulgaris*), *Seed Sci. Technol.,* 3, 719, 1975.

308. Ednie, A. B. and Needham, S. M., Laboratory test for internally-borne *Xanthomonas phaseoli* and *Xanthomonas phaseoli* var. *fuscans* in field bean (*Phaseolus vulgaris* L.) seed, *Proc. Assoc. Off. Seed Anal.,* 63, 76, 1973.

309. Grogan, R. G., Control of lettuce mosaic with virus free seed, *Plant Dis.,* 64, 446, 1980.

310. Grogan, R. C., Welch, J. E., and Bardin, R., Common lettuce mosaic and its control by the use of mosaic free seed, *Phytopathology,* 42, 573, 1952.

311. Kimble, K. A., Grogan, R. G., Greathead, A. S., Paulus, A. O., and House, J. K., Development, application and comparison of methods for indexing lettuce seed for mosaic virus in California, *Plant Dis. Rep.,* 59, 461, 1975.

312. Marrou, J., Messiaen, C. M., and Migliori, A., Méthode de controle de l'etat sanitaire des graines de laitue, *Ann. Epiphyt.,* 18, 227, 1967.

313. Carroll, T. W., Economic importance and control of barley stripe mosaic virus in Montana, in 3rd Int. Congr. Plant Pathology, P. Parey, Berlin, 1978, 30.

314. Carroll, T. W., Certification schemes against barley stripe mosaic, *Seed Sci. Technol.,* 11, 1033, 1983.

315. Hampton, R. O., Mink, G. I., Hamilton, R. I., Kraft, J. M., and Meuhlbauer, F. J., Occurrence of pea seedborne mosaic virus in North American Pea breeding lines, and procedures for its elimination, *Plant. Dis. Rep.,* 60, 455, 1976.

316. Bos, L. and VanderWant, J. P. H., Early browning of pea, a disease caused by a soil and seedborne virus, *Tijdschr. Plantenziekten,* 68, 368, 1962.

317. Polston, J. E. and Goodman, R. M., Enzyme linked immunosorbent assay (ELISA) to produce virus free plants from soybean germplasm, *Phytopathology,* 71, 250, 1981.

318. Neergaard, P., The infection percentage as a relative value in assessing disease tolerance for seed health testing, *Proc. Int. Seed Test. Assoc.,* 27, 400, 1962.

319. Walker, J. C. and Patel, P. N., Splash dispersal and wind as factors in epidemiology of halo blight on bean, *Phytopathology,* 54, 140, 1964.

320. Zink, F. W., Grogan, R. G., and Wetch, J. E., The effect of the percentage of seed transmission upon subsequent spread of lettuce mosaic virus, *Phytopathology,* 46, 662, 1956.

321. Baker, K. F., Seed pathology, in *Seed Biology,* Vol. 2, Kozlowski, T. T., Ed., Academic Press, New York, 1972, 317.

322. Morton, D. J., A quick method of preparing barley embryos for loose smut examination, *Phytopathology,* 50, 270, 1960.

323. Agarwal, V. K., Singh, O. V., and Singh, A., A note on certification standard for the karnal bunt disease of wheat, *Seed Res.,* 1, 97, 1973.

324. Agarwal, V. K., Singh, A., and Verma, H. S., Outbreak of karnal bunt of wheat, *FAO Plant Prot. Bull.,* 24, 99, 1976.

325. Mathys, G. and Baker, E. A., An appraisal of the effectiveness of quarantines, *Annu. Rev. Phytopathol.,* 18, 85, 1980.

326. Kahn, R. P., Plant quarantine: principles, methodology and suggested approaches, in *Plant Health and Quarantine in International Transfer of Genetic Resources,* Hewitt, W. B. and Chiarappa, L., Eds., CRC Press, Cleveland, 1977, 289.

327. Waterworth, H. E. and White, G. A., Plant introductions and quarantine: the need for both, *Plant Dis.,* 66, 87, 1982.

328. Neergaard, P., A review on quarantine for seed, in *Golden Jubilee Commemoration Volume,* National Academy of Sciences, New Dehli, 1980.

329. Peregrine, W. T. H., Groundnut rust (*Puccinia arachidis*) in Brunei, *PANS,* 17, 318, 1971.

330. Mendez, M., Quarantine initiatives on seed pathology in the Americas, *Seed Pathol. News,* 9, 1, 1976.

331. West, E., Peanut rust, *Plant Dis. Rep.,* 15, 5, 1931.

332. Neergaard, P., Seed health — policy of certification and disease control, *Seed Pathol. News,* 6, 7, 1974.

333. Chiarappa, L., Man-made epidemiological hazards in major crops of developing countries, in *Plant Diseases and Vectors Ecology and Epidemiology,* Maramorosch, K. and Harris, K. F., Eds., Academic Press, New York, 1981, 319.

334. Bingefers, S., International dispersal of nematodes, *Neth. J. Plant Pathol.,* 73 (Suppl. 1), 44, 1967.

335. Chock, A. K., The international plant protection convention, in *Plant Health, the Scientific Basis for Administrative Control of Plant Diseases and Pests,* Ebbels, D. L. and King, J. E., Eds., Blackwell Scientific, Oxford, 1979, 1.

336. Singh, K. G., Regional ASEAN collaborations in plant quarantine, *Seed Sci. Technol.,* 11, 1189, 1983.

337. Hodge, W. H. and Erlanson, C. O., Federal plant introduction — a review, *Econ. Bot.,* 10, 299, 1956.

338. Rohwer, G. G., Plant quarantine philosophy of the United States, in *Plant Health, the Scientific Basis for Administrative Control of Plant Diseases and Pests,* Ebbels, D. L. and King, J. E., Eds., Blackwell Scientific, Oxford, 1980, 23.

339. Leppik, E. E., Introduced seed-borne pathogens endanger crop breeding and plant introduction, *FAO Plant Prot. Bull.,* 16, 57, 1968.

340. Ebbels, D. L. and King, J. E., Eds., *Plant Health, the Scientific Basis for Administrative Control of Plant Diseases and Pests,* Blackwell Scientific, Oxford, 1979, 322.

341. Southey, J. F., Preventing the entry of alien diseases and pests into Great Britain, in *Plant Health, the Scientific Basis for Administrative Control of Plant Diseases and Pests*, Ebbels, D. L. and King, J. E., Eds., Blackwell Scientific, Oxford, 1979, 63.

342. Wadhi, S. R., Plant Quarantine Activity at the National Bureau of Plant Genetic Resources, NBPGR Sci. Mongr. No. 2, National Bureau of Plant Genetic Resources, New Delhi, 1980, 99.

343. Olembo, S., Seed health testing at the plant quarantine station at Muguga, Kenya, *Seed Sci. Technol.*, 11, 1217, 1983.

344. Morrison, L. G., *Quarantine Principles and Policy, SPC Workshop and Training Course in Plant Quarantine, Suva, Fiji*, SPC, Suva, 1977, 2.

345. Sethi, C. L., Nath, R. P., Mathur, V. K., and Ahuja, S., Interceptions of plant parasitic nematodes from imported seed/plant material, *Indian J. Nematol.*, 2, 89, 1972.

346. Sheffield, F. M. L., Requirements of a post-entry quarantine station, *FAO Plant Prot. Bull.*, 6, 149, 1958.

347. Sheffield, F. M. L., Closed quarantine procedures, *Rev. Appl. Mycol.*, 47, 1, 1968.

348. Anon., *Plant Pathologist's Pocketbook*, Commonwealth Mycological Institute, Kew, England, 1974, 267.

349. Berg, G. H., Post-entry and intermediate quarantine stations, in *Plant Health and Quarantine in International Transfer of Genetic Resources*, Hewitt, W. B. and Chiarappa, L., Eds., CRC Press, Cleveland, 1977, 315.

350. Leppik, E. E., Seed-borne Diseases on Introduced Plants, in Rep. No. 7, North Central Regional Plant Introduction Station, Ames, Ia., 1962, 11.

351. Leppik, E. E., List of foreign pests, pathogens and weeds detected on introduced plants, in Plant Introduction Investigation Paper No. 15, U.S. Department of Agriculture, Beltsville, Md., 1969.

352. Tisselli, O., Sinclair, J. B., and Hymowitz, T., Sources of Resistance to Selected Fungal, Bacterial, Viral and Nematode Diseases of Soybeans, INTSOY Ser. No. 10, College of Agriculture, University of Illinois at Urbana-Champaign, 1980, 134.

353. Hooker, A. L., Southern leaf blight of corn — present status and future prospects, *J. Environ. Qual.*, 1, 244, 1972.

354. Champion, R., Testing cultivars for resistance to disease in France, *Seed Sci. Technol.*, 11, 681, 1983.

355. Prasad, K. and Weigle, J. L., Association of seed coat factors with resistance to *Rhizoctonia solani* in *Phaseolus vulgaris*, *Phytopathology*, 66, 342, 1976.

356. Braverman, S. W., Aseptic culture of soybean and peanut embryonic axes to improve phytosanitation of plant introductions, *Seed Sci. Technol.*, 3, 725, 1975.

INDEX

T

Printed and bound by CPI Group (UK) Ltd, Croydon, CR0 4YY

22/10/2024

01777638-0014